An Analogue of a Reductive Algebraic Monoid Whose Unit Group Is a Kac-Moody Group

MEMOIRS
of the
American Mathematical Society

Number 823

An Analogue of a Reductive Algebraic Monoid Whose Unit Group Is a Kac-Moody Group

Claus Mokler

March 2005 • Volume 174 • Number 823 (third of 4 numbers) • ISSN 0065-9266

American Mathematical Society
Providence, Rhode Island

2000 *Mathematics Subject Classification.*
Primary 17B67, 22E65.

Library of Congress Cataloging-in-Publication Data

Mokler, Claus, 1962–
An analogue of a reductive algebraic monoid whose unit group is a Kac-Moody group / Claus Mokler.
p. cm. — (Memoirs of the American Mathematical Society, ISSN 0065-9266 ; no. 823)
"Volume 174, number 823 (third of 4 numbers)."
Includes bibliographical references.
ISBN 0-8218-3648-X (alk. paper)
1. Kac-Moody algebras. 2. Lie groups. 3. Algebroids. I. Title. II. Series.

QA3.A57 no. 823
[QA252.3]
510 s—dc22
[512′.55] 2004062690

Memoirs of the American Mathematical Society

This journal is devoted entirely to research in pure and applied mathematics.

Subscription information. The 2005 subscription begins with volume 173 and consists of six mailings, each containing one or more numbers. Subscription prices for 2005 are $606 list, $485 institutional member. A late charge of 10% of the subscription price will be imposed on orders received from nonmembers after January 1 of the subscription year. Subscribers outside the United States and India must pay a postage surcharge of $31; subscribers in India must pay a postage surcharge of $43. Expedited delivery to destinations in North America $35; elsewhere $130. Each number may be ordered separately; *please specify number* when ordering an individual number. For prices and titles of recently released numbers, see the New Publications sections of the *Notices of the American Mathematical Society.*

Back number information. For back issues see the *AMS Catalog of Publications.*

Subscriptions and orders should be addressed to the American Mathematical Society, P. O. Box 845904, Boston, MA 02284-5904, USA. *All orders must be accompanied by payment.* Other correspondence should be addressed to 201 Charles Street, Providence, RI 02904-2294, USA.

Memoirs of the American Mathematical Society is published bimonthly (each volume consisting usually of more than one number) by the American Mathematical Society at 201 Charles Street, Providence, RI 02904-2294, USA. Periodicals postage paid at Providence, RI. Postmaster: Send address changes to Memoirs, American Mathematical Society, 201 Charles Street, Providence, RI 02904-2294, USA.

This publication is indexed in *Science Citation Index®*, *SciSearch®*, *Research Alert®*, *CompuMath Citation Index®*, *Current Contents®/Physical, Chemical & Earth Sciences.*
Printed in the United States of America.

∞ The paper used in this book is acid-free and falls within the guidelines established to ensure permanence and durability.
Visit the AMS home page at `http://www.ams.org/`

10 9 8 7 6 5 4 3 2 1 10 09 08 07 06 05

Contents

Abstract

By an easy generalization of the Tannaka-Krein reconstruction we associate to the category of admissible representations of the category $\mathcal{O}$ of a Kac-Moody algebra, and its category of admissible duals, a monoid with a coordinate ring.

The Kac-Moody group is the Zariski open dense unit group of this monoid. The restriction of the coordinate ring to the Kac-Moody group is the algebra of strongly regular functions introduced by V. Kac and D. Peterson.

This monoid has similar structural properties as a reductive algebraic monoid. In particular it is unit regular, its idempotents related to the faces of the Tits cone. It has Bruhat and Birkhoff decompositions.

The Kac-Moody algebra is isomorphic to the Lie algebra of this monoid.

Received by the editor June 5, 2002, and in revised form January 16, 2004.

2000 *Mathematics Subject Classification.* 17B67, 22E65.

Key words and phrases. Kac-Moody groups, algebra of strongly regular functions, reductive algebraic monoids.

Introduction

The (minimal) Kac-Moody group G constructed in [**K,P 1**] by V. Kac and D. Peterson is a group analogue of a semisimple, simply connected algebraic group (without a coordinate ring, without a topology). V. Kac and D. Peterson defined and investigated in [**K,P 2**] the algebra of strongly regular functions $\mathbb{C}[G]$ of a symmetrizable Kac-Moody group. This algebra has many properties in common with the coordinate ring of a semisimple, simply connected algebraic group. It is an integrally closed domain, even a unique factorization domain. It admits a Peter and Weyl theorem, i.e.,

$$\mathbb{C}[G] \cong \bigoplus_{\Lambda \in P^+} L^*(\Lambda) \otimes L(\Lambda)$$

as $G \times G$-modules. But the following things, which hold in the non-classical case, are different:

- The group structure of G does not induce a Hopf algebra structure on $\mathbb{C}[G]$. There is no comultiplication due to the infinite-dimensionality of the highest and lowest weight representations. There is no antipode due to the in-equivalence between highest and lowest weight representations.
- Assigning to every element of G its vanishing ideal, G embeds in the set of 1-codimensional ideals of $\mathbb{C}[G]$, which we denote by $\operatorname{Specm}\mathbb{C}[G]$. But this map is not surjective.

V. Kac and D. Peterson posed the problem to determine the part

$$(0.1) \qquad \bigcup_{m \in \mathbb{N}} \ \bigcup_{\beta_1, \ldots, \beta_m \in \Delta_{re}} \overline{U_{\beta_1} \cdots U_{\beta_m} T}^{Specm} \subseteq \operatorname{Specm}\mathbb{C}[G] \ .$$

(Here the elements of G have been identified with the corresponding vanishing ideals. T denotes a maximal torus given by the construction of the Kac-Moody group, and $U_{\beta_1}, \ldots, U_{\beta_m}$ are root groups belonging to T.)

At the same time P. Slodowy wanted to connect the deformation theory of certain singularities with the Kac-Moody groups. Together with E. Looijenga he constructed an adjoint quotient

$$\chi : G \ \rightarrow \ \widehat{T}/\mathcal{W}$$

as a map, [**Sl 1**], [**Sl 2**]. In the non-classical case the target space $\widehat{T}/\mathcal{W}$ is a proper extension of $T/\mathcal{W}$, related to a certain partial compactification described in [**Loo**]. (Such an extension is really necessary. This is indicated by the fact, that not every element of G can be conjugated in a fixed Borel subgroup.) Compared to its classical counterpart this adjoint quotient has bad properties. In the non-classical case there are 'missing' conjucagy classes in its fibres. P. Slodowy guessed that this quotient can be extended to a certain part $\widehat{G} \subseteq \operatorname{Specm} \mathbb{C}[G]$. There is an action of $G \times G$ on $\operatorname{Specm} \mathbb{C}[G]$, extending the action on G. P. Slodowy conjectured

$$\widehat{G} = G \cdot \widehat{T} \cdot G \,, \tag{0.2}$$

where $\widehat{T}$ should be realized as a torus embedding $T \subseteq \widehat{T}$ of non-finite type, [**Sl 2**]. He also guessed the form of certain elements $e(R(\Theta)) \in \widehat{T}$, Θ special.

He discussed this with D. Peterson. Some time later D. Peterson announced a proof of this conjecture, $\widehat{G}$ the part (0.1) of $\operatorname{Specm} \mathbb{C}[G]$, together with some other structural properties of $\widehat{G}$, [**P**]. The most important are:

- There should be a monoid structure on $\widehat{G}$.
- Every idempotent of $\widehat{G}$ is conjugate to one of the elements $e(R(\Theta))$ by an element of G.
- $\widehat{G} = U_\pm \widehat{N} U_\pm$, where $\widehat{N} := N\{\ e(R(\Theta)) \mid \Theta \text{ special }\ \}N$ is a submonoid of $\widehat{G}$.

But D. Peterson didn't give any proof.

About ten years later, as a student of P. Slodowy, the task of my PhD was to investigate all these conjectures and claims, [**M 1**].

Important for these investigations was the paper [**Kas**] of M. Kashiwara, in which he had studied the whole spectrum $\operatorname{Spec} \mathbb{C}[G]$ in connection with his infinite dimensional algebraic geometric approach to the flag varieties of Kac-Moody groups. For his construction of the flag variety he used an open subscheme of the full spectrum $\operatorname{Spec} \mathbb{C}[G]$, which has a countable covering by big cells. He factored this subscheme by an action of the formal Borel subgroup of the formal Kac-Moody group G_f.

There had also been another development relevant for these investigations: A reductive linear algebraic monoid $(M, \mathbb{C}[M])$ is a connected linear algebraic monoid, whose principal open unit group G is reductive. The linear algebraic monoid $M(n, \mathbb{C})$ of $n \times n$-matrices is a typical example. Note that there are no non-trivial reductive linear algebraic monoids with semisimple unit group.

Reductive linear algebraic monoids have been studied carefully by M. Putcha and L. Renner. Many of their results are presented in the books [**Pu**], [**Re 2**]. Such a monoid M is unit regular. Every idempotent of M is G-conjugate to an idempotent in the closure $\overline{T}$ of a maximal torus T of G. It has Bruhat and Birkhoff decompositions, the Weyl group N/T replaced by the Renner monoid $\overline{N}/T$. The closure $\overline{T}$ of the torus T is an affine toric variety. The Weyl group $\mathcal{W}$ acts on $\overline{T}$, extending its action on T. L. Renner showed in [**Re 1**] that a normal reductive linear algebraic monoid M with zero and one-dimensional center is classified by its unit group G, and the toric variety $\overline{T}$, equipped with the Weyl group action. The same holds if the assumption of a one-dimensional center is dropped, [**Re 2**]. E. Vinberg gave in [**Vi**] a representation theoretic approach to the classification problem.

To investigate the part (0.1) of $\operatorname{Specm} \mathbb{C}[G]$, which we call Kac-Peterson-Slodowy part, to prove the conjectures of P. Slodowy, and the claims of D. Peterson we shall proceed as follows:

Since the Kac-Moody group G acts faithfully on the sum $\bigoplus_{\Lambda\in P^+} L(\Lambda)$, it can be identified with with a subgroup of the monoid of linear endomorphisms $End(\bigoplus_{\Lambda\in P^+} L(\Lambda))$. In Chapter 1 we define a monoid $\widehat{G}$, which is generated by G and certain idempotents $e(R(\Theta)) \in End(\bigoplus_{\Lambda\in P^+} L(\Lambda))$, Θ special. For this monoid we prove the conjecture (0.2) of P. Slodowy, and all the claims of D. Peterson on the structure. We obtain even more results. We show

$$\widehat{G} \;=\; \dot{\bigcup_{\hat{\sigma}\in\widehat{\mathcal{W}}}} B^{\pm}\hat{\sigma}B^{\pm} \;,$$

and investigate the monoid $\widehat{\mathcal{W}} = \widehat{N}/T$, which we call Weyl monoid. We prove a theorem, which allows to reduce any computation in $\widehat{G}$ to computations in the Kac-Moody group G, and in the face lattice $\mathcal{R}(X)$ of the Tits cone X, equipped with its action of the Weyl group. We show how to construct the monoid $\widehat{G}$ from the twin root datum $(G, (U_\alpha)_{\alpha\in\Delta_{re}})$ and from the monoid $(\,\mathcal{R}(X)\,,\,\cap\,)$.

It is not too difficult to see, that $\widehat{G}$ embeds in Specm $\mathbb{C}\,[G]$ in a natural way, its image contained in the Kac-Peterson-Slodowy part. But I could not find a direct proof, showing equality. Also it is surprising that there exists such a monoid. It is desirable to gain a better understanding where this monoid comes from, and to make the analogy with a reductive algebraic monoid more plain.

A semisimple, simply connected algebraic group can be obtained from its category of rational representations by a Tannaka-Krein reconstruction. It is also possible to use for this reconstruction the corresponding category of representations of its semisimple Lie algebra. Because we assumed 'simply connected', this category coincides with the category of finite dimensional representations of the semisimple Lie algebra.

It is important to observe, that there are several natural generalizations of this category for a Kac-Moody algebra **g**:

- The tensor category of admissible modules of **g**. (These modules are used for the construction of the Kac-Moody group G. The condition 'admissible' is slightly stronger as the more common 'integrable'. It guarantees that the maximal torus T is big enough.)
- The tensor category $\mathcal{O}_{adm}$ of admissible modules of the category $\mathcal{O}$. This category is close to the classical case, because it keeps the complete reducibility theorem, i.e., every module of $\mathcal{O}_{adm}$ is isomorphic to a direct sum of irreducible highest weight modules $L(\Lambda)$, $\Lambda \in P^+$.
- A category, which is build in a similar way as $\mathcal{O}_{adm}$, but where we have a complete reducibility theorem using the lowest weight modules $L^*(\Lambda)$, $\Lambda \in P^+$.

In a way similar to the Tannaka-Krein reconstruction, a suitable category of representations of a Lie algebra, together with a suitable category of duals determine a monoid with coordinate ring. This monoid is the biggest monoid, acting reasonably on the representations, compatible with the duals. The coordinate ring is a coordinate ring of matrix coefficients. But the multiplication of the monoid does not have to induce a comultiplication of the coordinate ring.

In Chapter 4 we formulate this generalized Tannaka-Krein reconstruction for the category $\mathcal{O}_{adm}$ and its category of admissible duals. In Chapter 5 we show that $\widehat{G}$ is the monoid determined by these categories, which is one of the main results of this book. The Kac-Moody group G is the Zariski open dense unit group of $\widehat{G}$. The algebra of strongly regular functions $\mathbb{C}[G]$ is isomorphic to the coordinate ring $\mathbb{C}[\widehat{G}]$ by the restriction map.

In this way, the algebra of strongly regular functions is really the coordinate ring of a monoid. Note that this monoid is a purely infinite-dimensional phenomenon. In the classical case, it reduces to a semisimple, simply connected algebraic group.

It is quite a long way to prove this result. We use an easy infinite-dimensional algebraic geometry, which is developed in Chapter 3. The monoids obtained by a generalized Tannaka-Krein reconstruction are weak algebraic monoids in this algebraic geometric setting.

As remarked above, it is not too difficult to see, that the part of $\mathrm{Specm}\,\mathbb{C}[G]$ corresponding to $\widehat{G}$ is contained in the Kac-Peterson-Slodowy part. Using the characterization of $\widehat{G}$ by the generalized Tannaka-Krein reconstruction, we show in Chapter 5, that the Kac-Peterson-Slodowy part is contained in the part of $\mathrm{Specm}\,\mathbb{C}[G]$ corresponding to $\widehat{G}$. Therefore we have equality.

Furthermore we show in Chapter 5, that the Zariski closure $\overline{T} = \widehat{T}$ is an affine toric variety of non-finite type. We show $\overline{N} = \widehat{N}$, which makes the analogy of the Weyl monoid $\widehat{\mathcal{W}} = \widehat{N}/T = \overline{N}/T$ with a Renner monoid even more plain. Although the monoid $\widehat{G}$ does not act on the flag varieties, we show that it acts on the affine cones of the flag varieties.

In Chapter 6 we show that the Kac-Moody algebra is isomorphic to the Lie algebra of $\widehat{G}$.

In Chapter 1 we quickly review Kac-Moody algebras, Kac-Moody groups and the algebra of strongly regular functions. We also give an easy generalization of affine toric varieties.

The monoid $\widehat{G}$ can be characterized as the closure of the Kac-Moody group G in a very big space. To find and to investigate this closure can be motivated by the approach of E. Vinberg to the classification problem of reductive algebraic monoids. I found the interpretation by the generalized Tannaka-Krein reconstruction after my PhD. I included it here, because it shows the nature of this monoid much better.

Many other things in relation to this monoid have already been investigated: The $\mathbb{C}$-valued points of the algebra of strongly regular functions have been determined and investigated in [**M 2**]. The results on the $G_f \times G_f$-orbits of $\mathrm{Specm}\,\mathbb{C}[G]$ (closures, big cells, stratified transversal slices) can be transfered to the $G\times G$-orbits of $\widehat{G}$, for which they are easier to prove.

The Bruhat and Birkhoff cells of $\widehat{G}$ have been investigated in [**M 3**], in particular their closures have been determined.

The generalized Tannaka-Krein reconstruction has been investigated as a way to integrate infinite-dimensional Lie algebras. It was natural to determine the monoid, which corresponds to the category $\mathcal{O}_{adm}$ and its category of full duals. This is a monoid $\widehat{G_f}$, which is build in a similar way as the monoid $\widehat{G}$, but the minimal Kac-Moody group G is replaced by the formal Kac-Moody group G_f, [**M 4**].

Working with similar categories of lowest weight modules of the Kac-Moody algebra we get monoids $\check{G}$ and $\check{G}_f$ which are isomorphic as monoids with coordinate rings to $\widehat{G}$ and $\widehat{G_f}$. Presumably the Kac-Moody group G is the monoid associated to the category of admissible modules and its category of admissible duals. This is difficult to show, because there is no complete reducibility theorem.

The adjoint quotient map $\chi : G \to \widehat{T}/\mathcal{W}$ of E. Looijenga and P. Slodowy can be entirely defined by using the monoid $\widehat{G}$. It looks as if this map really extends to a map $\chi : \widehat{G} \to \widehat{T}/\mathcal{W}$, whose fibres have better properties. It is a striking difference to the classical case that for understanding the conjucagy classes of the group G one has to complete G to a monoid $\widehat{G}$.

CHAPTER 1

Preliminaries

In the first section we collect the main facts about Kac-Moody algebras, Kac-Moody groups, and the algebra of strongly regular functions, which are used later. We also introduce the notation used in this book.

In the second section we collect some results on certain generalizations of affine toric varieties, whose coordinate rings don't need to be finitely generated.

1.1. Kac-Moody algebras, Kac-Moody groups and the algebra of strongly regular functions

The Kac-Moody group given in [**K,P 1**], [**K,P 3**] corresponds to the derived Kac-Moody algebra. We work with a slightly enlarged group corresponding to the full Kac-Moody algebra as in [**Ti 1**], [**Mo,Pi**].

All the material stated in this subsection about Kac-Moody algebras can be found in the books [**K**] (most results also valid for a field of characteristic zero, with the same proofs), [**Mo,Pi**], about Kac-Moody groups in [**K,P 1**], [**K,P 3**], [**Mo,Pi**], and about the algebra of strongly regular functions in [**K,P 2**].

We denote by $\mathbb{N} = \mathbb{Z}^+$, $\mathbb{Q}^+$, respectively $\mathbb{R}^+$ the sets of strictly positive numbers of $\mathbb{Z}$, $\mathbb{Q}$, respectively $\mathbb{R}$, and the sets $\mathbb{N}_0 = \mathbb{Z}_0^+$, $\mathbb{Q}_0^+$, $\mathbb{R}_0^+$ contain in addition the zero.

In the whole book $\mathbb{F}$ is a field of characteristic 0, and $\mathbb{K}$ an arbitrary field with $|\mathbb{K}| = \infty$. Their unit groups are denoted by $\mathbb{F}^\times$ and $\mathbb{K}^\times$.

Generalized Cartan matrices: Starting point for the construction of a Kac-Moody algebra and its associated simply connected Kac-Moody group is a *generalized Cartan matrix*, which is a matrix $A = (a_{ij}) \in M_n(\mathbb{Z})$ with $a_{ii} = 2$, $a_{ij} \leq 0$ for all $i \neq j$, and $a_{ij} = 0$ if and only if $a_{ji} = 0$. Denote by l the rank of A and set $I := \{1, 2, \ldots, n\}$. For the properties of the generalized Cartan matrices, in particular their classification, we refer to the book [**K**]. In the whole book we assume A to be symmetrizable.

Optimal realizations: A simply connected minimal free root base for A, which we call an *optimal realization* for short, consists of dual free $\mathbb{Z}$-modules H, P of rank $2n - l$, and linear independent sets $\Pi^\vee = \{h_1, \ldots, h_n\} \subseteq H$, $\Pi = \{\alpha_1, \ldots, \alpha_n\} \subseteq P$, such that $\alpha_i(h_j) = a_{ji}$, $i, j = 1, \ldots, n$. Furthermore, $H_I := Q^\vee := \mathbb{Z}\text{-span}\{h_1, \ldots, h_n\}$ is saturated in H, which means that for all $n \in \mathbb{N}$, $x \in H$ we have: $\quad nx \in H_I \;\Rightarrow\; x \in H_I$.

P is called the *weight lattice*, $Q := \mathbb{Z}\text{-span}\{\ \alpha_i \mid i \in I\ \}$ the *root lattice*, and $Q^\vee$ the *coroot lattice*. Set $Q_0^\pm := \mathbb{Z}_0^\pm\text{-span}\{\ \alpha_i \mid i \in I\ \}$, and $Q^\pm := Q_0^\pm \setminus \{0\}$.

We fix a complement H_{rest} of H_I in H. This complement determines a system of *fundamental dominant weights* $\Lambda_1, \ldots, \Lambda_n$ by

$$\Lambda_i(h_j) := \delta_{ij} \quad (j = 1, \ldots, n) \quad , \quad \Lambda_i(h) := 0 \quad (h \in H_{rest}) \ .$$

We extend $h_1, \ldots, h_n \in H_I$ with elements $h_{n+1}, \ldots, h_{2n-l} \in H_{rest}$ to a base of H, and extend $\Lambda_1, \ldots, \Lambda_n$ to the corresponding dual base $\Lambda_1, \ldots, \Lambda_{2n-l}$.

The Weyl group and the Tits cone: Identify H and P with the corresponding lattices in the $\mathbb{F}$-linear spaces

$$\mathbf{h} := \mathbf{h}_{\mathbb{F}} := H \otimes_{\mathbb{Z}} \mathbb{F} \quad , \quad \mathbf{h}^* := \mathbf{h}^*_{\mathbb{F}} := P \otimes_{\mathbb{Z}} \mathbb{F} \ .$$

$\mathbf{h}^*$ is interpreted as the dual of $\mathbf{h}$. Order the elements of $\mathbf{h}^*$ by $\lambda \leq \lambda'$ if and only if $\lambda' - \lambda \in Q_0^+$.

Choose a symmetric matrix $B \in M_n(\mathbb{Q})$, a diagonal matrix $D = \operatorname{diag}(\epsilon_1, \ldots, \epsilon_n)$, $\epsilon_1, \ldots, \epsilon_n \in \mathbb{Q}^+$, such that $A = DB$. Define a nondegenerate symmetric bilinear form on $\mathbf{h}$ by

$$\begin{aligned} (h_i \mid h) = (h \mid h_i) &:= \alpha_i(h)\,\epsilon_i & & i \in I, \quad h \in \mathbf{h}\,, \\ (h' \mid h'') &:= 0 & & h', h'' \in \mathbf{h}_{rest} := H_{rest} \otimes \mathbb{F}\ . \end{aligned}$$

Denote by $\nu : \mathbf{h} \to \mathbf{h}^*$ the corresponding linear isomorphism. Denote by $(\mid)$ the induced nondegenerate symmetric form on $\mathbf{h}^*$.

The *Weyl group* $\mathcal{W} = \mathcal{W}(A)$ is the Coxeter group with generators σ_i, $i \in I$, and relations

$$\sigma_i^2 = 1 \quad (i \in I) \quad , \quad (\sigma_i\sigma_j)^{m_{ij}} = 1 \quad (i, j \in I,\, i \neq j) \ .$$

The numbers m_{ij} are given by:

$a_{ij}a_{ji}$	0	1	2	3	≥ 4
m_{ij}	2	3	4	6	no relation between σ_i and σ_j

The Weyl group $\mathcal{W}$ acts faithfully and contragrediently on $\mathbf{h}$ and $\mathbf{h}^*$ by

$$\begin{aligned} \sigma_i h &:= h - \alpha_i(h) h_i \quad , \quad i \in I, \quad h \in \mathbf{h}, \\ \sigma_i \lambda &:= \lambda - \lambda(h_i)\,\alpha_i \quad , \quad i \in I, \quad \lambda \in \mathbf{h}^*, \end{aligned}$$

leaving the lattices $Q^\vee$, H, Q, P, and the forms invariant. The set $\Delta_{re} := \mathcal{W}\{\, \alpha_i \mid i \in I \,\} \subseteq Q$ is called the set of *real roots*, and $\Delta_{re}^\vee := \mathcal{W}\{\, h_i \mid i \in I \,\} \subseteq Q^\vee$ the set of *real coroots*. The map $\alpha_i \mapsto h_i$, $i \in I$, can be extended to a $\mathcal{W}$-equivariant bijection $\alpha \mapsto h_\alpha = \frac{2\nu^{-1}(\alpha)}{(\alpha|\alpha)}$ of Δ_{re} and $\Delta_{re}^\vee$.

To illustrate the action of $\mathcal{W}$ on $\mathbf{h}^*_{\mathbb{R}}$ geometrically, for $J \subseteq I$ set

$$\begin{aligned} F_J &:= \{\, \lambda \in \mathbf{h}^*_{\mathbb{R}} \mid \lambda(h_i) = 0 \text{ for } i \in J, \ \lambda(h_i) > 0 \text{ for } i \in I \setminus J \,\} \ , \\ \overline{F}_J &:= \{\, \lambda \in \mathbf{h}^*_{\mathbb{R}} \mid \lambda(h_i) = 0 \text{ for } i \in J, \ \lambda(h_i) \geq 0 \text{ for } i \in I \setminus J \,\} \ . \end{aligned}$$

$\overline{F}_J$ is a closed, finitely generated, convex cone with relative interior F_J, and

$$\overline{F}_J = \dot{\bigcup_{I \supseteq K \supseteq J}} F_K \ .$$

We call $\overline{C} := \{\, \lambda \in \mathbf{h}^*_{\mathbb{R}} \mid \lambda(h_i) \geq 0 \text{ for } i \in I \,\}$ the *(closed) fundamental chamber*, and $C := \{\, \lambda \in \mathbf{h}^*_{\mathbb{R}} \mid \lambda(h_i) > 0 \text{ for } i \in I \,\}$ the *open fundamental chamber*. The *Tits cone*

$$X := \mathcal{W}\overline{C}$$

is a convex $\mathcal{W}$-invariant cone with edge $c := \{ \lambda \in \mathbf{h}_{\mathbb{R}}^* \mid \lambda(h_i) = 0 \text{ for } i \in I \}$, and fundamental region $\overline{C} = \bigcup_{J \subseteq I} F_J$. The parabolic subgroup $\mathcal{W}_J$ of $\mathcal{W}$ is the stabilizer of any element $\lambda \in F_J$, as well as the stabilizer of F_J as a whole. The set $\{ \sigma F_J \mid \sigma \in \mathcal{W},\ J \subseteq I \}$ gives a $\mathcal{W}$-invariant partition of X. We call σF_J a *facet* of *type* J.

We call the weights of the set $P^+ := P \cap \overline{C}$ the *dominant* weights of the weight lattice P, and the weights of the set $P^{++} := P \cap C$ the *regular dominant* weights of the weight lattice P.

The Kac-Moody algebra: The *Kac-Moody algebra* $\mathbf{g} = \mathbf{g}(A)$ is the Lie algebra over $\mathbb{F}$ generated by the abelian Lie algebra $\mathbf{h}$, and $2n$ elements e_i, f_i, $(i \in I)$, with the following relations, which hold for any $i, j \in I$, $h \in \mathbf{h}$:

$$[e_i, f_j] = \delta_{ij} h_i \ , \quad [h, e_i] = \alpha_i(h) e_i \ , \quad [h, f_i] = -\alpha_i(h) f_i \ ,$$
$$(ad\, e_i)^{1-a_{ij}} e_j = (ad\, f_i)^{1-a_{ij}} f_j = 0 \qquad (i \neq j) \ .$$

The abelian Lie algebra $\mathbf{h}$ and the elements e_i, f_i, $(i \in I)$, can be identified with their images in $\mathbf{g}$.

The *Chevalley involution* $*$ of $\mathbf{g}$ is the involutive anti-automorphism determined by $e_i^* = f_i$, $f_i^* = e_i$, $h^* = h$, $(i \in I$ and $h \in \mathbf{h})$.

The nondegenerate symmetric bilinear form $(\ \mid\ )$ on $\mathbf{h}$ extends uniquely to a nondegenerate symmetric invariant bilinear form $(\ \mid\ )$ on $\mathbf{g}$.

There is the *root space decomposition*

$$\mathbf{g} = \bigoplus_{\alpha \in \mathbf{h}^*} \mathbf{g}_\alpha \quad \text{where} \quad \mathbf{g}_\alpha := \{ x \in \mathbf{g} \mid [h, x] = \alpha(h) x \ \text{ for all } \ h \in \mathbf{h} \} \ .$$

In particular, $\mathbf{g}_0 = \mathbf{h}$, $\mathbf{g}_{\alpha_i} = \mathbb{F} e_i$, and $\mathbf{g}_{-\alpha_i} = \mathbb{F} f_i$, $i \in I$.

The set of *roots* $\Delta := \{ \alpha \in \mathbf{h}^* \setminus \{0\} \mid \mathbf{g}_\alpha \neq 0 \}$ is invariant under the Weyl group and spans the root lattice Q. We have $\Delta_{re} \subseteq \Delta$, and $\Delta_{im} := \Delta \setminus \Delta_{re}$ is called the set of *imaginary roots.* Δ, Δ_{re}, and Δ_{im} decompose into the disjoint union of the sets of *positive* and *negative roots* $\Delta^\pm := \Delta \cap Q^\pm$, ${\Delta_{re}}^\pm := \Delta_{re} \cap Q^\pm$, ${\Delta_{im}}^\pm := \Delta_{im} \cap Q^\pm$. We have $\Delta^\pm = -\Delta^\mp$, ${\Delta_{re}}^\pm = -{\Delta_{re}}^\mp$, ${\Delta_{im}}^\pm = -{\Delta_{im}}^\mp$.

The decomposition $\mathbf{g} = \mathbf{n}^- \oplus \mathbf{h} \oplus \mathbf{n}^+$, where $\mathbf{n}^\pm := \bigoplus_{\alpha \in \Delta^\pm} \mathbf{g}_\alpha$, is called the *triangular decomposition.*

For a real root α the subalgebra $\mathbf{s}_\alpha := \mathbf{g}_\alpha \oplus [\mathbf{g}_\alpha, \mathbf{g}_{-\alpha}] \oplus \mathbf{g}_{-\alpha}$ of $\mathbf{g}$ is isomorphic to $sl(2, \mathbb{F})$.

The derived Lie algebra of $\mathbf{g}$ is given by $\mathbf{g}' = \bigoplus_{\alpha \in \Delta} \mathbf{g}_\alpha \oplus \mathbf{h}'$, where $\mathbf{h}'$ is spanned by the elements h_i, $i \in I$.

The category $\mathcal{O}$: The objects of the category $\mathcal{O}$ consist of the $\mathbf{g}$-modules V with the following properties:

(1) V is $\mathbf{h}$-diagonalizable with finite-dimensional weight spaces.
(2) There exist finitely many elements $\lambda_1, \ldots, \lambda_m \in \mathbf{h}^*$, such that the set of weights of V, which we denote by $P(V)$, is contained in the union $\bigcup_{1=1}^m D(\lambda_i)$, where $D(\lambda_i) := \{ \lambda \in \mathbf{h}^* \mid \lambda \leq \lambda_i \}$.

The morphisms of $\mathcal{O}$ are morphisms of $\mathbf{g}$-modules.

The irreducible highest weight representations: For every $\Lambda \in \mathbf{h}^*$ there exists, unique up to isomorphism, an irreducible representation $(L(\Lambda), \pi_\Lambda)$ of $\mathbf{g}$, with a nonzero element v_Λ satisfying

$$\pi_\Lambda(\mathbf{n}^+)v_\Lambda = 0 \quad , \quad \pi_\Lambda(h)v_\Lambda = \Lambda(h)v_\Lambda \quad (h \in \mathbf{h}) .$$

It is is an object of $\mathcal{O}$, and is called an irreducible highest weight representation. We denote its set of weights by $P(\Lambda)$.

Any such representation carries a nondegenerate symmetric bilinear form $\langle\langle \mid \rangle\rangle : L(\Lambda) \times L(\Lambda) \to \mathbb{F}$, which is contravariant, i.e. $\langle\langle v \mid xw \rangle\rangle = \langle\langle x^*v \mid w \rangle\rangle$ for all $v, w \in L(\Lambda)$, $x \in \mathbf{g}$. This form is unique up to a nonzero multiplicative scalar.

The Kac-Moody group: To construct the Kac-Moody group call a representation (V, π) of $\mathbf{g}$ *admissible* if:

(1) V is $\mathbf{h}$-diagonalizable with set of weights $P(V) \subseteq P$.
(2) $\pi(x)$ is locally nilpotent for all $x \in \mathbf{g}_\alpha$, $\alpha \in \Delta_{re}$.

(Note: If the generalized Cartan matrix is degenerate, then this definition is slightly stronger than 'integrable', because we demand $P(V) \subseteq P$ in addition. This guarantees that we can integrate the full Cartan subalgebra $\mathbf{h}$ to an algebraic geometric torus.)

Examples are the irreducible highest weight representations $(L(\Lambda),\ \pi_\Lambda)$, $\Lambda \in P^+$, and the adjoint representation $(\mathbf{g}\,,\ ad\,)$.

We denote by $\mathcal{O}_{adm}$ the full subcategory of the category $\mathcal{O}$, whose objects are admissible modules. Every object of this category is isomorphic to a direct sum of irreducible highest weight modules $L(\Lambda)$, $\Lambda \in P^+$.

Let $\tilde{G}$ be the free product of the additive groups $\mathbf{g}_\alpha$, $\alpha \in \Delta_{re}$, and the torus $\mathrm{Hom}((P, +), (\mathbb{F}^\times, \cdot))$. Let (V, π) be an admissible representation. Due to (1), (2), and the coproduct property of $\tilde{G}$ we get a homomorphism $\tilde{\pi} : \tilde{G} \to GL(V)$, mapping $x_\alpha \in \mathbf{g}_\alpha$ to $\exp(\pi(x_\alpha))$, and $\epsilon \in \mathrm{Hom}(P, \mathbb{F}^\times)$ to an element $t_\pi(\epsilon)$, defined by

$$t_\pi(\epsilon)v_\lambda := \epsilon(\lambda)v_\lambda \quad , \quad v_\lambda \in V_\lambda \, , \, \lambda \in P(V) .$$

Let $\tilde{N}$ be the intersection of all kernels of homomorphisms $\tilde{\pi}'$ associated to admissible representations (V, π'). The *Kac-Moody group* is defined as $G := G(A) := \tilde{G}/\tilde{N}$. Due to its definition $\tilde{\pi} : \tilde{G} \to GL(V)$ factors to a representation $\Pi : G \to GL(V)$. Corresponding to $(\mathbf{g}\,,\ ad\,)$ we get the adjoint representation $Ad : G \to Aut(\mathbf{g})$.

By composing the injection of $\mathbf{g}_\alpha$ into $\tilde{G}$ with the projection onto G we get an injective homomorphism $\exp : \mathbf{g}_\alpha \to G$. Its image U_α is called the *root group* belonging to $\alpha \in \Delta_{re}$.

Similarly we get an injective homomorphism $t : \mathrm{Hom}(P, \mathbb{F}^\times) \to G$, its image denoted by T. Identifying $H \otimes_{\mathbb{Z}} \mathbb{F}^\times$ with $\mathrm{Hom}(P, \mathbb{F}^\times)$, the torus T can be described by the isomorphism

$$\begin{array}{ccc} H \otimes_{\mathbb{Z}} \mathbb{F}^\times & \to & T \\ \sum_\tau h_\tau \otimes s_\tau & \mapsto & \prod_\tau t_{h_\tau}(s_\tau) \end{array} ,$$

where

$$t_h(s)v_\lambda := s^{\lambda(h)}v_\lambda, \quad v_\lambda \in V_\lambda, \quad \lambda \in P(V)$$

Set $t_i(s) := t_{h_i}(s)$, $i = 1, \ldots, 2n - l$, $s \in \mathbb{F}^\times$.

The derived group G', which is identical with the Kac-Moody group as defined in [**K,P 1**], is generated by the root groups U_α, $\alpha \in \Delta_{re}$. We have $G = G' \rtimes T_{rest}$, where T_{rest} is the subtorus of T generated by the elements $t_i(s)$, $i = n+1, \ldots, 2n-l$, $s \in \mathbb{F}^\times$.

The Kac-Moody group acts faithfully on the sum $\bigoplus_{\Lambda \in P^+} L(\Lambda)$ and also on the sum $\bigoplus_{i=1,\ldots,2n-l} L(\Lambda_i)$.

The Chevalley involution $* : G \to G$ is the involutive anti-isomorphism determined by

$$\exp(x_\alpha)^* := \exp(x_\alpha^*) \quad (x_\alpha \in \mathbf{g}_\alpha\,,\ \alpha \in \Delta_{re}) \quad , \quad t^* := t \quad (t \in T)\,.$$

It is compatible with any nondegenerate symmetric contravariant form $\langle\langle \mid \rangle\rangle$ on any of the modules $L(\Lambda)$, $\Lambda \in P^+$, i.e., $\langle\langle v \mid gw \rangle\rangle = \langle\langle g^* v \mid w \rangle\rangle$ for all $v, w \in L(\Lambda)$, $g \in G$.

The Kac-Moody group has the following important structural properties:

(1) Let $\alpha \in \Delta_{re}^+$ and $x_\alpha \in \mathbf{g}_\alpha$, $x_{-\alpha} \in \mathbf{g}_{-\alpha}$, such that $[x_\alpha, x_{-\alpha}] = h_\alpha$. There exists an injective homomorphism of groups $\phi_\alpha : \mathrm{SL}(2, \mathbb{F}) \to G$ with

$$\phi_\alpha \begin{pmatrix} 1 & s \\ 0 & 1 \end{pmatrix} = \exp(s x_\alpha) \quad , \quad \phi_\alpha \begin{pmatrix} 1 & 0 \\ s & 1 \end{pmatrix} = \exp(s x_{-\alpha}) \quad , \quad s \in \mathbb{F}^\times .$$

Its image is denoted by G_α.

(2) For $s \in \mathbb{F}^\times$ set

$$n_\alpha(s) := \phi_\alpha \begin{pmatrix} 0 & s \\ -\frac{1}{s} & 0 \end{pmatrix},$$

and for short set $n_i(s) := n_{\alpha_i}(s)$, $i = 1, \ldots, n$. Note that

$$t_{h_\alpha}(s) = \phi_\alpha \begin{pmatrix} s & 0 \\ 0 & \frac{1}{s} \end{pmatrix} = n_\alpha(s) n_\alpha(1)^{-1}.$$

Denote by N the subgroup generated by T and $n_\alpha(1)$, $\alpha \in \Delta_{re}$. The Weyl group $\mathcal{W}$ can be identified with the group N/T by the isomorphism $\kappa : \mathcal{W} \to N/T$ given by $\kappa(\sigma_\alpha) := n_\alpha(1)T$, $\alpha \in \Delta_{re}$. We denote an arbitrary element $n \in N$ with $\kappa^{-1}(nT) = \sigma \in \mathcal{W}$ by n_σ.

If (V, π) is an admissible $\mathbf{g}$-module, its set of weights $P(V)$ is $\mathcal{W}$-invariant, and $n_\sigma V_\lambda = V_{\sigma\lambda}$, $\lambda \in P(V)$. We have $n_\sigma U_\alpha n_\sigma^{-1} = U_{\sigma\alpha}$, $\alpha \in \Delta_{re}$.

(3) Denote by $B^\pm$ the subgroups of G generated by the torus T and the root groups U_α, $\alpha \in \Delta_{re}^\pm$.

The system $(\,G\,,\,(U_\alpha)_{\alpha \in \Delta_{re}})$ is a twin root datum, compare [**K,P 3**], Proposition 4.7, and [**Ti 2**], with associated twin BN-pairs $B^\pm$, N. These satisfy $B^+ \cap B^- = B^+ \cap N = B^- \cap N = T$. The common Coxeter group is N/T. There are the Bruhat and Birkhoff decompositions

$$G = \dot{\bigcup_{\sigma \in \mathcal{W}}} B^\epsilon \sigma B^\delta \quad , \quad (\epsilon, \delta) = \underbrace{(+,+)\,,\,(-,-)}_{Bruhat}\,,\,\underbrace{(+,-)\,,\,(-,+)}_{Birkhoff}\,.$$

(4) Let $U^\pm$ be the subgroups of G generated by the root groups U_α, $\alpha \in \Delta_{re}^\pm$. The standard parabolic subgroups

$$P_J^\pm = B^\pm \mathcal{W}_J B^\pm \quad , \quad J \subseteq I\,,$$

admit the Levi decompositions

$$P_J^\pm = \tilde{G}_J \ltimes (U^J)^\pm\,.$$

Here $\tilde{G}_J$ is the group generated by the groups $U_{\pm\alpha_j}$, $j \in J$, and by T. This group coincides with the intersection $P_J^+ \cap P_J^-$. The groups $(U^J)^\pm$ are the normal subgroup of $U^\pm$, or of $(P_J)^\pm$, generated by U_α, $\alpha \in \Delta_{re}^\pm \setminus \sum_{j\in J} \mathbb{Z}\alpha_j$. They coincide with $\bigcap_{\sigma\in\mathcal{W}_J} \sigma U^\pm \sigma^{-1}$. In particular,

$$B^\pm \;=\; T \ltimes U^\pm \,.$$

Usually the upper indices + of all these groups are omitted.

We set $\tilde{N}_J := \tilde{G}_J \cap N$. This group is generated by the elements $n_j(1)$, $j \in J$, and by T.

The algebra of strongly regular functions: Let $\Lambda \in P^+$. The integrable irreducible lowest weight module $L^*(\Lambda)$ of lowest weight $-\Lambda$ can be realized as the submodule $\bigoplus_{\lambda\in P(\Lambda)} (L(\Lambda)_\lambda)^*$ of the full dual module $L(\Lambda)^*$ of $L(\Lambda)$.

For $\phi \in L^*(\Lambda)$, and $v \in L(\Lambda)$ call the function $\tilde{f}_{\phi v} : G \to \mathbb{F}$ given by $\tilde{f}_{\phi v}(g) := \phi(gw)$, $g \in G$, a *matrix coefficient* of G.

The algebra $\mathbb{F}[G]$ generated by all such matrix coefficients is called the *algebra of strongly regular functions* on G. It is an integrally closed domain, and admits a *Peter and Weyl theorem*: Define an action of $G \times G$ on $\mathbb{F}[G]$ by

$$((g,h)f)(x) \;:=\; f(g^{-1}xh)\,, \qquad g,\, x,\, h \in G, \quad f \in \mathbb{F}[G]\,.$$

Then the map

$$\bigoplus_{\Lambda\in P^+} L^*(\Lambda) \otimes L(\Lambda) \;\to\; \mathbb{F}[G] \quad \text{induced by} \quad \phi \otimes v \mapsto \tilde{f}_{\phi v}$$

is an isomorphism of $G \times G$-modules.

There is an alternative, slightly different way to state this theorem: Define an action of $G \times G$ on $\mathbb{F}[G]$, and an involution $*$ of $\mathbb{F}[G]$, by

$$\begin{array}{rcl} ((g,h)f)(x) & := & f(g^*xh) \\ f^*(x) & := & f(x^*) \end{array}\,, \qquad g,\, x,\, h \in G, \quad f \in \mathbb{F}[G]\,.$$

For every $\Lambda \in P^+$ fix a nondegenerate symmetric contravariant bilinear form $\langle\langle\ \mid\ \rangle\rangle$ on the admissible irreducible highest weight module $L(\Lambda)$. For $v, w \in L(\Lambda)$ let $\tilde{f}_{vw} : G \to \mathbb{F}$ be the function defined by $\tilde{f}_{vw}(g) := \langle\langle v \mid gw \rangle\rangle$, $g \in G$. Then the map

$$\bigoplus_{\Lambda\in P^+} L(\Lambda) \otimes L(\Lambda) \;\to\; \mathbb{F}[G] \quad \text{induced by} \quad v \otimes w \mapsto \tilde{f}_{vw}$$

is an isomorphism of $G \times G$-modules. It identifies the sum of the switch-maps with the involution.

Restricting the functions of $\mathbb{F}[G]$ to G' respectively T_{rest} we get the algebras $\mathbb{F}[G']$ respectively $\mathbb{F}[T_{rest}]$. The first is identical with the algebra of strongly regular functions as defined in [**K,P 2**], the second is the classical coordinate ring of the torus T_{rest}. $\mathbb{F}[G]$ is isomorphic to $\mathbb{F}[G'] \otimes \mathbb{F}[T_{rest}]$ by the comorphism of the multiplication map $G' \times T_{rest} \to G$.

Substructures: For $\emptyset \neq J \subseteq I$ the submatrix $A_J := (a_{ij})_{i,j\in J}$ of A is a generalized Cartan matrix. There exist saturated sublattices $H(A_J) \subseteq H$, $P(A_J) \subseteq P$, with $(h_j)_{j\in J} \subseteq H(A_J)$, $(\alpha_j)_{j\in J} \subseteq P(A_J)$, giving an optimal realization for A_J. We have $P = P(A_J) \oplus H(A_J)^\perp$, and the projections of Λ_j, $j \in J$, onto $P(A_J)$ give a system of fundamental dominant weights of this optimal realization.

The corresponding Kac-Moody algebra $\mathbf{g}(A_J)$ embeds in $\mathbf{g}$. Identifying $\mathbf{g}(A_J)$ with its image, the root lattice $Q(A_J)$ identifies with the sublattice $Q_J := \sum_{j\in J}\mathbb{Z}\alpha_j$ of Q. The Weyl group $\mathcal{W}(A_J)$ identifies with the parabolic subgroup $\mathcal{W}_J$ of $\mathcal{W}$. The set of roots $\Delta(A_J)$ identifies with $\Delta_J := \Delta \cap Q_J$, similarly the set of real roots $\Delta(A_J)_{re}$ identifies with $(\Delta_J)_{re} := \Delta_{re} \cap Q_J = \mathcal{W}_J\{\alpha_j \mid j \in J\}$. (To simplify the notation we sometimes identify the set of roots $\{\alpha_j \mid j \in J\}$ with J.) The Kac-Moody group $G(A_J)$ embeds in G in the obvious way.

The images of $\mathbf{g}(A_J)$ and $G(A_J)$ under these embeddings depend on the choice of the sublattice $H(A_J)$. The images of $\mathbf{g}(A_J)'$ and $G(A_J)'$ are independent of this choice, because the sublattice $H_J := \mathbb{Z}\text{-}span\{\ h_j \mid j \in J\ \}$ of $H(A_J)$ is uniquely determined by A_J. We denote the image of $\mathbf{g}(A_J)'$, respectively $G(A_J)'$, by $\mathbf{g}_J$, respectively G_J. It would be more consequent to write $\mathbf{g}'_J$ and G'_J instead of $\mathbf{g}_J$ and G_J, but to simplify the notation we omit the prime. The coordinate ring $\mathbb{F}[G(A_J)']$ identifies with the restrictions of $\mathbb{F}[G]$ on G_J. (A similar statement for $\mathbb{F}[G(A_J)]$ is not valid.)

The root space decomposition and the triangular decomposition of $\mathbf{g}(A_J)$ induce a $Q(A_J)$-grading and a triangular decomposition of $\mathbf{g}(A_J)'$. Identifying $\mathbf{g}(A_J)'$ with $\mathbf{g}_J$ these are given by $\mathbf{g}_J = \mathbf{n}_J^- \oplus \mathbf{h}_J \oplus \mathbf{n}_J^+$ with $\mathbf{n}_J^\pm = \bigoplus_{\alpha\in\Delta_J^\pm}\mathbf{g}_\alpha \cong \mathbf{n}(A_J)^\pm$, and $\mathbf{h}_J$ spanned by the elements h_j, $j \in J$.

The subtorus $T_J := T \cap G_J$ of T is generated by the elements $t_j(s)$, $j \in J$, $s \in \mathbb{F}^\times$. The group $N_J := N \cap G_J$ is the subgroup of G generated by the elements $n_j(1)$, $j \in J$, and by T_J. The groups $U_J^\pm := U^\pm \cap G_J$ are generated by the root groups U_α, $\alpha \in (\Delta_J)_{re}^\pm$. They are isomorphic to $U(A_J)^\pm$. Set $B_J^\pm := B \cap G_J$. The twin root datum of $G(A_J)$ induces the twin root datum $(G_J\,,\,(U_\alpha)_{\alpha\in(\Delta_J)_{re}})$ with associated twin BN-pairs $B_J^\pm$, N_J. These satisfy $B_J \cap B_J^- = B_J \cap N_J = B_J^- \cap N_J = T_J$. The common Coxeter group N_J/T_J is isomorphic to $\mathcal{W}_J$. The corresponding Bruhat and Birkhoff decompositions of G_J are induced by the Bruhat and Birkhoff decompositions of G. There are Levi decompositions of the standard parabolic subgroups, in particular $B_J^\pm = T_J \ltimes U_J^\pm$.

To simplify the notation of many formulas, it is useful to set $\mathbf{g}_\emptyset := \mathbf{n}_\emptyset := \{0\}$, $G_\emptyset := T_\emptyset := U_\emptyset := N_\emptyset := \{1\}$, and $\mathbb{F}[G_\emptyset] := \mathbb{F}\,1$.

The case of a decomposable generalized Cartan matrix: If the generalized Cartan matrix is decomposable, $A = A_{I_1} \oplus A_{I_2}$, we can decompose the optimal realization and H_{rest}. Set

$$\begin{aligned} H_{rest\,1} &:= \{\ x \in H_{rest} \mid \alpha_i(x) = 0 \ \text{ for all } \ i \in I_2\ \}\ , \\ H_{rest\,2} &:= \{\ x \in H_{rest} \mid \alpha_i(x) = 0 \ \text{ for all } \ i \in I_1\ \}\ . \end{aligned}$$

We get optimal realizations of A_1 respectively A_2 by

$$\begin{aligned} H_1 &:= H_{I_1} \oplus H_{rest\,1} &,\quad \Pi_1^\vee &:= \{\ h_i \mid i \in I_1\ \}\ , \\ P_1 &:= \{\ \lambda \in P \mid \lambda(h) = 0 \ \text{ for } \ h \in H_2\ \} &,\quad \Pi_1 &:= \{\ \alpha_i \mid i \in I_1\ \}\ , \end{aligned}$$

respectively

$$\begin{aligned} H_2 &:= H_{I_2} \oplus H_{rest\,2} &,\quad \Pi_2^\vee &:= \{\ h_i \mid i \in I_2\ \}\ , \\ P_2 &:= \{\ \lambda \in P \mid \lambda(h) = 0 \ \text{ for } \ h \in H_1\ \} &,\quad \Pi_2 &:= \{\ \alpha_i \mid i \in I_2\ \}\ . \end{aligned}$$

These satisfy:

$$\begin{aligned} H &= H_1 \oplus H_2 &,\quad \Pi^\vee &= \Pi_1^\vee \,\dot\cup\, \Pi_2^\vee\ . \\ P &= P_1 \oplus P_2 &,\quad \Pi &= \Pi_1 \,\dot\cup\, \Pi_2\ . \end{aligned}$$

The Weyl group $\mathcal{W}(A_i)$ can be identified with the parabolic subgroup $\mathcal{W}_{I_i}$ of $\mathcal{W}$, $i = 1, 2$, and we have $\mathcal{W} = \mathcal{W}_{I_1} \times \mathcal{W}_{I_2}$. If X_{A_i} denotes the Tits cone of the optimal realization of A_i, $i = 1, 2$, then $X = X_{A_1} \oplus X_{A_2}$.

The Lie algebra $\mathbf{g}_i := \bigoplus_{\alpha \in \Delta_{I_i}} \mathbf{g}_\alpha \oplus \text{span}(H_i)$, the subgroup G_i generated by U_α, $\alpha \in (\Delta_{I_i})_{re}$, and by the elements t_h, $h \in H_i$, and the coordinate ring $\mathbb{F}[G_i]$ obtained by restricting the functions of $\mathbb{F}[G]$ onto G_i are in the obvious way isomorphic to $g(A_{I_i})$, $G(A_{I_i})$, and $\mathbb{F}[G(A_{I_i})]$, $i = 1, 2$. Furthermore we have $\mathbf{g} = \mathbf{g}_1 \oplus \mathbf{g}_2$, $G = G_1 \times G_2$, and $\mathbb{F}[G] = \mathbb{F}[G_1] \otimes \mathbb{F}[G_2]$.

1.2. A generalization of affine toric varieties

An affine toric variety is a normal affine variety containing a torus T as a dense open subset, together with an action of T on the variety, that extends the natural action of T on itself. Alternatively an affine toric variety can be described as a normal algebraic monoid, whose unit group is a torus T. It is determined by a rational convex polyhedral cone, and can be constructed, starting with this cone, [**Fu**], [**Ne**].

In this section we state a generalization of this construction, starting with a not necessarily finitely generated rational convex cone. The results are similar as in the classical case, but some of the classical proofs can not be generalized. The proofs of these results, as well as some other results, can be found in [**M 1**]. Here we omit the proofs, because certainly they can be also found elsewhere in the literature.

Rational convex cones and saturated submonoids of a lattice. Let L be a lattice of finite rank. Identify L with the corresponding lattice in the real vector space $V := L \otimes_{\mathbb{Z}} \mathbb{R}$.

Let $Y \subseteq V$. Denote by $\dot{c}c\,(Y)$ the pointed convex cone of V, denote by $cc\,(Y)$ the convex cone of V, and by $\text{span}\,(Y)$ the subspace of V, which is generated by Y. The property 'finitely generated' is defined in the obvious way.

Definition 1.1. A pointed convex cone, respectively a convex cone, respectively a subspace X of V is called rational, if there exists a subset of L which generates X.

A subsemigroup, respectively a submonoid, respectively a subgroup S of L is called saturated, if for all $x \in L$ and $n \in \mathbb{N}$ we have: $\quad nx \in S \quad \Rightarrow \quad x \in S$

The next theorem generalizes the Lemma of Gordan and its converse:

Theorem 1.2. *The following maps are inverse, and respect the property 'finitely generated':*

$$\begin{array}{ccc} \{\textit{ rational pointed convex cones of } V\,\} & & \{\textit{ saturated subsemigroups of } L\,\} \\ X & \mapsto & X \cap L \\ \dot{c}c\,(S) & \leftarrow\!\shortmid & S \end{array}$$

Similar things hold if

'rational pointed convex cones', 'saturated subsemigroups', and '$\dot{c}c(S)$'

are replaced by

'rational convex cones', 'saturated submonoids', and '$cc(S)$',

or if they are replaced by

'rational subspaces', 'saturated subgroups', and '$span(S)$'.

We call X and S associated, if they correspond to one another under these maps.

Let $X \subseteq V$ be a convex cone. Recall the definitions of the faces of X, the face lattice $Fa(X)$, the exposed faces of X, the relative interior of faces, and the hull of faces, [**Ro**]. We denote a convex cone of the form $X - F$, where F is a face of X, a *dual face* of X.

The algebraic forms of these definitions can be imitated for submonoids of a lattice: Let M be a submonoid of L.

A subset $F \subseteq M$ is called a *face* of M, if F is a submonoid, and $M \setminus F$ is a semigroup ideal, (i.e., $M \setminus F = \emptyset$, or else $M \setminus F \neq \emptyset$ and $M \setminus F + M \subseteq M \setminus F$). The set of faces of M, denoted by $Fa(M)$, together with the inclusion of faces, is a lattice. The intersection of faces coincides with the lattice intersection. If F is a face of M, then the set of faces of F is given by

$$Fa\,(F) \;=\; \{\; G \in Fa\,(M) \;\mid\; G \subseteq F \;\}\;.$$

Define the *relative interior* of a face F of M as

$$riF \;:=\; F \setminus \bigcup_{G \in Fa(F)\,,\, G \neq F} G\;.$$

The set $\{\; riF \;\mid\; F \in Fa\,(M)\;\}$ is a partition of M. The face F is the smallest face containing an element $m \in riF$.

The *hull* of a face F is the subgroup of L generated by F, which is equal to $F - F$. Note that $F = M \cap (F - F)$.

We denote a monoid $M - F$, where F is a face of M, a *dual face* of M. Its set of faces is given by

$$Fa\,(M-F) \;\;=\;\; \{\; G-F \mid G \in Fa\,(M) \text{ with } G \supseteq F\;\}\;.$$

Theorem 1.3. **(a)** *Let X be a rational convex cone of V. Every face, relative interior of a face, hull of a face, and dual face of X is rational.*

(b) *Let M be a saturated submonoid of L. Every face, relative interior of a face, hull of a face, and dual face of M is saturated.*

(c) *Let X and M be associated. By the maps corresponding to cones and submonoids of the last theorem, the face lattices $(Fa\,(X)\,,\subseteq)$ and $(Fa\,(M)\,,\subseteq)$ are isomorphic. The relative interiors, hulls, and dual faces of associated faces are associated.*

Our main example will be the Tits cone X, which is a P-rational convex cone in $\mathbf{h}_{\mathbb{R}}^*$. Its faces will be given explicitely. Because of their facetial structures, it is also not difficult to verify the last theorem directly.

Generalized affine toric varieties. Let M be a saturated submonoid of a lattice of finite rank. Recall that $\mathbb{K}$ denotes a field with $|\mathbb{K}| = \infty$. The set of homomorphisms of monoids

$$\widetilde{M} := \operatorname{Hom}((M,+),(\mathbb{K},\cdot))$$

gets the structure of an abelian monoid by multiplying the homomorphisms pointwise.

We identify the monoid algebra $\mathbb{K}[M]$ with a coordinate ring $\mathbb{K}[\widetilde{M}]$ on $\widetilde{M}$, identifying $\sum_{m\in M} c_m m \in \mathbb{K}[M]$ (which means $c_m \in \mathbb{K}$, $c_m \neq 0$ only for finitely many $m \in M$) with the function

$$\left(\sum_{m\in M} c_m\, m\right)(\alpha) := \sum_{m\in M} c_m\, \alpha(m) \quad , \quad \alpha \in \widetilde{M}\,.$$

The monoid structure of $\widetilde{M}$ induces a coalgebra structure on $\mathbb{K}[\widetilde{M}]$. We equip $\widetilde{M}$ with the Zariski topology induced by its coordinate ring. $\widetilde{M}$ is irreducible and normal, i.e., the coordinate ring $\mathbb{K}[\widetilde{M}]$ is normal.

$\widetilde{M}$ maps bijectively to the $\mathbb{K}$-valued points of $\mathbb{K}[\widetilde{M}] = \mathbb{K}[M]$, the inverse map given by restricting the $\mathbb{K}$-valued points onto $M \subseteq \mathbb{K}[M]$.

Note that if M is a subgroup, then $\widetilde{M} = \operatorname{Hom}(M,\mathbb{K}) = \operatorname{Hom}(M,\mathbb{K}^\times)$ is a torus.

Proposition 1.4. *Let M be a saturated submonoid of a lattice of finite rank. For $F \in Fa(M)$ set*

$$T(F) := \left\{ \alpha \in \widetilde{M} \;\middle|\; \alpha^{-1}(\mathbb{K}^\times) = F \right\}\,.$$

Let $e(F) \in T(F)$ be the element given by

$$e(F)m := \begin{cases} 1 & \text{if } m \in F \\ 0 & \text{if } m \in M \setminus F \end{cases}\,.$$

Then we have:

(1) *$T(M)$ is the unit group of $\widetilde{M}$, and $E := \{\, e(F) \mid F \in Fa(M) \,\}$ is the set of idempotents of $\widetilde{M}$. Furthermore, $\widetilde{M} = T(M)\,E = E\,T(M)$.*

(2) *$T(F)$ is a subgroup of $\widetilde{M}$ with unit $e(F)$, isomorphic to the torus $\widetilde{F-F}$, an isomorphism $\Phi_F : \widetilde{F-F} \to T(F)$ given by*

$$\Phi_F(\alpha)(m) := \begin{cases} \alpha(m) & m \in F \\ 0 & m \in M \setminus F \end{cases} \quad , \quad \alpha \in \widetilde{F-F}\,.$$

Let the torus $T(M)$ act on $\widetilde{M}$ by left multiplication. The tori $T(F)$, $F \in Fa(M)$, are the $T(M)$-orbits of $\widetilde{M}$.

(3) *For $m \in M$ denote by $D(m)$ the principal open set $\{\alpha \in \widetilde{M} \mid \alpha(m) \neq 0\}$. We have*

$$\overline{T(F)} = \bigcup_{G\in Fa(F)} T(G)\,,$$

$$D(m) = \bigcup_{G\in Fa(M)\,,\,G\supseteq F} T(G) \quad \text{where} \quad m \in ri\,F\,.$$

In particular, $T(M)$ is principal open and dense in $\widetilde{M}$.

Remark 1.5. To use later, note that we get a surjective homomorphism of groups $\Psi_F : \widetilde{M - M} \to T(F)$ by

$$\Psi_F(\alpha)(m) := \begin{cases} \alpha(m) & m \in F \\ 0 & m \in M \setminus F \end{cases} \quad , \qquad \alpha \in \widetilde{M - M} .$$

CHAPTER 2

The monoid $\widehat{G}$ and its structure

Since the Kac-Moody group G acts faithfully on the sum $\bigoplus_{\Lambda\in P^+} L(\Lambda)$, it can be identified with a subgroup of the monoid of endomorphisms $End\left(\bigoplus_{\Lambda\in P^+} L(\Lambda)\right)$. The monoid $\widehat{G} \subseteq End\left(\bigoplus_{\Lambda\in P^+} L(\Lambda)\right)$ is defined to be generated by the Kac-Moody group $G \subseteq End\left(\bigoplus_{\Lambda\in P^+} L(\Lambda)\right)$, which is its unit group, and certain projections associated to the faces of the Tits cone.

We show some formulas which allow to reduce any computation in $\widehat{G}$ to computations in G and in the face lattice of the Tits cone equipped with its Weyl group action. We use these formulas to investigate the monoid $\widehat{G}$. We show that $\widehat{G}$ is unit regular. We give some double coset partitions of $\widehat{G}$, in particular the Bruhat and Birkhoff decompositions. For this reason we introduce the Weyl monoid. We give a representation independent description of $\widehat{G}$. It uses only the twin root datum $(G,\ (U_\alpha)_{\alpha\in\Delta re})$, and the face lattice of the Tits cone equipped with its Weyl group action.

We show that $\widehat{G}$ acts on the modules belonging to the category $\mathcal{O}_{adm}$, compatible with morphisms, and compatible with the tensor product of modules.

We give some results concerning submatrices of the generalized Cartan matrix and decomposable generalized Cartan matrices.

From now on for the whole book we fix for every $\Lambda \in P^+$ a highest weight module $L(\Lambda)$ with a nondegenerate contravariant symmetric bilinear from $\langle\langle\ \mid\ \rangle\rangle$. We extend to a form on the sum $\bigoplus_{\Lambda\in P^+} L(\Lambda)$, also denoted by $\langle\langle\ \mid\ \rangle\rangle$, by requiring $L(\Lambda)$ and $L(\Lambda')$ to be orthogonal for $\Lambda \neq \Lambda'$.

2.1. The face lattice of the Tits cone

Induced by the action on the Tits cone X, the Weyl group $\mathcal{W}$ acts on the face lattice of X, which we denote by $(\mathcal{R}(X), \subseteq)$. In this section we describe $(\mathcal{R}(X), \subseteq)$ together with its $\mathcal{W}$-action.

For $\emptyset \neq \Theta \subseteq I$ we denote by Θ^0, respectively Θ^∞ the set of indices corresponding to the sum of the components of A_Θ of finite, respectively non-finite type. We set $\emptyset^0 := \emptyset^\infty := \emptyset$.

Definition 2.1. A subset $\Theta \subseteq I$ is called special if $\Theta = \Theta^\infty$.

The special subsets of I are in one-to-one correspondence to certain faces of the Tits cone described in the following theorem. Part (1a), (1c), and (2) of this theorem are due to E. Looijenga, [**Loo**], Lemma 2.2, and Corollary 2.3. Part (1b) has been given by P. Slodowy in [**Sl 1**], Kapitel 6.2, Korollar 2. An alternative proof can be found in [**M 1**].

For $J \subseteq I$ set $J^\perp := \{\, i \in I \mid a_{ij} = 0 \text{ for all } j \in J \,\}$. Recall from the classification of the generalized Cartan matrices, that a set $\emptyset \neq \Theta \subseteq I$ is special if and only if $\left(\sum_{i\in\Theta} \mathbb{N}\, h_i\right) \cap (-\overline{C}^\vee) \neq \emptyset$.

Theorem 2.2. *Let Θ be special and set $R(\Theta) := \mathcal{W}_{\Theta^\perp}\overline{F}_\Theta$. Then:*

(1a) *$R(\Theta)$ is an exposed face of X, i.e.: We have $R(\emptyset) = X$. If $\Theta \neq \emptyset$ then for every element $c \in \left(\sum_{i\in\Theta} \mathbb{N}\, h_i\right) \cap (-\overline{C}^\vee)$ we have*

$$\begin{aligned} R(\Theta) &= X \cap \{\, \lambda \in \boldsymbol{h}_{\mathbb{R}}^* \mid \lambda(c) = 0 \,\} \;, \\ X &\subseteq \{\, \lambda \in \boldsymbol{h}_{\mathbb{R}}^* \mid \lambda(c) \geq 0 \,\} \;. \end{aligned}$$

(1b) *The relative interior of $R(\Theta)$ is given by*

$$ri\, R(\Theta) = \mathcal{W}_{\Theta^\perp} \bigcup_{\Theta^f \subseteq \Theta^\perp,\, \Theta^f = (\Theta^f)^0} F_{\Theta\cup\Theta^f} \;.$$

(1c) *The linear hull of $R(\Theta)$ is given by*

$$span(R(\Theta)) = \{\, \lambda \in \boldsymbol{h}_{\mathbb{R}}^* \mid \lambda(h_i) = 0 \;\; \text{for all} \;\; i \in \Theta \,\} \;.$$

(2) *The $\mathcal{W}$-stabilizers of $R(\Theta)$, pointwise and as a whole, are given by*

$$\begin{aligned} Z_{\mathcal{W}}(R(\Theta)) &:= \{\, \sigma \in \mathcal{W} \mid \sigma\lambda = \lambda \text{ for all } \lambda \in R(\Theta) \,\} = \mathcal{W}_\Theta \;, \\ N_{\mathcal{W}}(R(\Theta)) &:= \{\, \sigma \in \mathcal{W} \mid \sigma R(\Theta) = R(\Theta) \,\} = \mathcal{W}_{\Theta\cup\Theta^\perp} \;. \end{aligned}$$

The next two corollaries follow easily from Part (1b) of the last theorem. The first corollary relates the faces described in the last theorem to the closed fundamental chamber of the Tits cone. The second corollary, which has been already given by P. Slodowy in [**Sl 1**], Kapitel 6.2, Korollar 3, shows that these faces represent the $\mathcal{W}$-orbits of $\mathcal{R}(X)$.

Corollary 2.3. *A face R of the Tits cone X is of the form $R(\Theta)$, Θ special, if and only if $ri\, R \cap \overline{C} \neq \emptyset$.*

Corollary 2.4. *Every face of X is $\mathcal{W}$-conjugate to exactly one of the faces $R(\Theta)$, Θ special.*

Due to the last corollary, every face R can be parametrized in the form $R = \sigma R(\Theta)$ with $\sigma \in \mathcal{W}$, and with an uniquely determined special set Θ, which we call the *type* of R. The element σ is uniquely determined, if we restrict to the minimal coset representatives $\mathcal{W}^{\Theta\cup\Theta^\perp}$ of $\mathcal{W}/\mathcal{W}_{\Theta\cup\Theta^\perp}$.

The next proposition shows how to rewrite the inclusion of faces into conditions involving the parametrizations of the faces.

Proposition 2.5. *Let $\sigma, \sigma' \in \mathcal{W}$, Θ, Θ' be special. The following statements are equivalent:*

(i) $\sigma' R(\Theta') \subseteq \sigma R(\Theta)$

(ii) $\Theta' \supseteq \Theta$ *and* $\sigma^{-1}\sigma' \in \mathcal{W}_{\Theta^\perp}\mathcal{W}_{\Theta'}$

PROOF. '$(i) \Rightarrow (ii)$': Because $\sigma R(\Theta)$ contains facets of type Θ', we conclude $\Theta' \supseteq \Theta$. Comparing the facets of type Θ', we find $\sigma' F_{\Theta'} \subseteq \sigma \mathcal{W}_{\Theta^\perp} F_{\Theta'}$, which is equivalent to $\sigma^{-1}\sigma' \in \mathcal{W}_{\Theta^\perp}\mathcal{W}_{\Theta'}$.

'(ii) $\Rightarrow$ (i)': There exist $\tau_1 \in \mathcal{W}_{\Theta^\perp}$, $\tau_2 \in \mathcal{W}_{\Theta'}$, such that $\sigma^{-1}\sigma' = \tau_1\tau_2$. Because of $\Theta' \supseteq \Theta$, we have $\mathcal{W}_{\Theta'^\perp} \subseteq \mathcal{W}_{\Theta^\perp}$ and $\overline{F}_{\Theta'} \subseteq \overline{F}_\Theta$. Therefore

$$\sigma^{-1}\sigma' R(\Theta') = \tau_1 \mathcal{W}_{\Theta'^\perp}\overline{F}_{\Theta'} \subseteq \mathcal{W}_{\Theta^\perp}\overline{F}_\Theta = R(\Theta) .$$

□

It turns out to be difficult to rewrite the intersection of faces into conditions involving the parametrizations. But there is the following easy case: If Θ, Θ' are special, then also $\Theta \cup \Theta'$ is special, and we have $R(\Theta) \cap R(\Theta') = R(\Theta \cup \Theta')$.

Examples 2.6. The following examples describe the faces of the Tits cone for certain generalized Cartan matrices A.

(1) If A is of finite type the Tits cone is a linear space. We have $\mathcal{R}(X) = \{X\}$.

(2) If A is of affine or strongly hyperbolic type, then $\mathcal{R}(X) = \{c,\, X\}$, where c is the edge of the Tits cone.

(3) If A is of indefinite type, but not strongly hyperbolic, then it is not difficult to see, that there exist infinitely many faces. As an example consider the hyperbolic generalized Cartan matrix

$$A = \begin{pmatrix} 2 & -2 & 0 \\ -2 & 2 & -1 \\ 0 & -1 & 2 \end{pmatrix} .$$

The form $(\ |\ )$ restricted to $\mathbf{h}_\mathbb{R}^*$ has signature (1,2). The Tits cone X consists of one component of the open cone $\{ \lambda \in \mathbf{h}_\mathbb{R}^* \mid (\lambda \mid \lambda) < 0 \}$ together with all closed, P-rational, isotropic half-lines on the boundary, compare [**F,F**], [**Sl 2**], or [**M 1**]. The special sets are $\{1,2,3\}$, $\{1,2\}$, and $\emptyset$. The corresponding faces of X of these types are the edge $c = \{0\}$, every closed, P-rational, isotropic half-line on the boundary, and X itself.

The following two propositions concerning submatrices of the generalized Cartan matrix and decomposable generalized Cartan matrices are not difficult to prove. Use the theorem, the corollary, and the proposition of above.

Proposition 2.7. *Let $\emptyset \neq J \subseteq I$. Identify the Weyl group $\mathcal{W}(A_J)$ with the parabolic subgroup $\mathcal{W}_J$ of $\mathcal{W}$. The map*

$$\begin{aligned} \Upsilon_J :\ \mathcal{R}(X_{A_J}) &\to \mathcal{R}(X) \\ \tau R_{A_J}(\Theta) &\mapsto \tau R(\Theta) \end{aligned}$$

is an embedding of lattices, and has the following image:

$$\begin{aligned} \mathcal{R}(X)_J &:= \{ R \in \mathcal{R}(X) \mid R \supseteq R(J^\infty) \} \\ &= \{ \tau R(\Theta) \mid \Theta \subseteq J \text{ special},\ \tau \in \mathcal{W}_J \} \\ &= \{ \tau R(\Theta) \mid \Theta \subseteq J^\infty \text{ special},\ \tau \in \mathcal{W}_{J^\infty} \} . \end{aligned}$$

To simplify the notation of many formulas it is useful to set $\mathcal{R}(X)_\emptyset := \{X\}$.

Proposition 2.8. *Let A be decomposable, $A = A_1 \oplus A_2$. With the notions of Section 1.1, the embeddings Υ_{I_1}, Υ_{I_2} are given by*

$$\begin{aligned} \Upsilon_{I_1} :\ \mathcal{R}(X_{A_1}) &\to \mathcal{R}(X) \\ R &\mapsto R + X_{A_2} \end{aligned} \quad , \qquad \begin{aligned} \Upsilon_{I_2} :\ \mathcal{R}(X_{A_2}) &\to \mathcal{R}(X) \\ R &\mapsto X_{A_1} + R \end{aligned} \quad ,$$

and $\mathcal{R}(X)$ is the direct product of the sublattices $\mathcal{R}(X)_{I_1}$ and $\mathcal{R}(X)_{I_2}$.

2.2. The definition of the monoid $\widehat{G}$

The Tits cone X is P-rational. Recall the relation between the faces of X and the faces of the saturated submonoid $X \cap P$ of P, described in Section 1.2. Now the sets of weights $P(\Lambda)$ of the admissible highest weight modules $L(\Lambda)$, $\Lambda \in P^+$, satisfy

$$\bigcup_{\Lambda \in P^+} P(\Lambda) \;=\; X \cap P.$$

Definition 2.9. For a face R of the Tits cone X define a projection operator

$$e(R) \in End\left(\bigoplus_{\Lambda \in P^+} L(\Lambda) \right)$$

by $\quad e(R)v_\lambda := \begin{cases} v_\lambda & \lambda \in R \cap P \\ 0 & \lambda \notin R \cap P \end{cases} \quad , \quad v_\lambda \in L(\Lambda)_\lambda,\ \lambda \in P(\Lambda),\ \Lambda \in P^+.$

Denote its image by $B(R)$, and its kernel by $K(R)$. Set

$$E := \{\ e(R) \mid R \text{ a face of } X\ \} \quad , \quad E_{sp} := \{\ e(R(\Theta)) \mid \Theta \text{ special}\ \}\ .$$

Remark. Let $\Lambda \in F_J \cap P$, $J \subseteq I$. It is not difficult to check that $e(R)$ does not act as zero on $L(\Lambda)$ if and only if the type of R is contained in J.

The following proposition can be proved easily:

Proposition 2.10. **(1)** *E is in the obvious way a monoid isomorphic to the monoid $(\mathcal{R}(X), \cap)$.*

(2) *The monoid N acts by conjugation on E. Explicitely, let R be a face of the Tits cone X, and $n_\sigma \in N$. Then*

$$n_\sigma e(R) n_\sigma^{-1} \;=\; e(\sigma R)\ .$$

Furthermore, $n_\sigma B(R) = B(\sigma R)$ and $n_\sigma K(R) = K(\sigma R)$.

The Kac-Moody group G acts faithfully on $\bigoplus_{\Lambda \in P^+} L(\Lambda)$. We identify G with the corresponding subgroup of the monoid of endomorphisms $End\left(\bigoplus_{\Lambda\in P^+} L(\Lambda)\right)$.

Definition 2.11. $\widehat{G}$ is the submonoid of $End\left(\bigoplus_{\Lambda\in P^+} L(\Lambda)\right)$ generated by G and E.

Remark. The elements of E are defined with respect to the Cartan subalgebra $\mathbf{h}$ of $\mathbf{g}$. The monoid $\widehat{G}$ does not depend on this choice: Because G acts transitively on the Cartan subalgebras by the adjoint action, the set of projection operators defined with respect to $g\mathbf{h}$ is gEg^{-1}, $(g \in G)$.

Examples. There are the following easy cases, which can be checked directly from the definitions:

(1) If A is of finite type, then $\widehat{G} = G$.

(2) If A is of affine type, then $\widehat{G} = (G' \cup \{e(c)\}) \rtimes T_{rest}$. The element $e(c)$ is the zero of the monoid $G' \cup \{e(c)\}$.

(3) If A is of strongly hyperbolic type, then $\widehat{G} = G \cup \{e(c)\}$. The element $e(c)$ is the zero of the monoid $\widehat{G}$.

If A is of indefinite type and not strongly hyperbolic, then due to the infinitely many faces of the Tits cone, the monoid $\widehat{G}$ is much more complicated.

Recall that we fixed nondegenerate contravariant symmetric bilinear forms on the modules $L(\Lambda)$, $\Lambda \in P^+$, and extended these forms canonically to a form on the sum $\bigoplus_{\Lambda \in P^+} L(\Lambda)$. Taking the adjoint of an element of G is possible and coincides with taking the Chevalley involution, since the forms on the modules $L(\Lambda)$, $\Lambda \in P^+$, are contravariant. It is easy to check, that the projections $e(R)$, $R \in \mathcal{R}(X)$, are selfadjoint. Therefore taking the adjoint gives an involution of $\widehat{G}$.

Definition 2.12. We call this involution the Chevalley involution $*$ of $\widehat{G}$.

2.3. Formulas for computations in $\widehat{G}$

The elements of $\widehat{G}$ are given by expressions of the form

$$g_1 e(R_1) \cdots g_m e(R_m) g_{m+1} \quad \text{where} \quad \begin{array}{c} g_1, \ldots, g_{m+1} \in G \\ R_1, \ldots, R_m \in \mathcal{R}(X) \end{array}, \quad m \in \mathbb{N} . \tag{2.1}$$

It would be bad to investigate the structure of $\widehat{G}$ working with expressions of long length m. The first thing to do is to look for some formulas, which allow to reduce the length of such an expression (m=1 can be reached), and which allow to decide if two such expressions give the same element. These are given in Theorem 2.20, which is in essential of the following form:

Let $g, g' \in G$ and $R, R' \in \mathcal{R}(X)$. Then:

$$ge(R) = e(R')g' \quad \Longleftrightarrow \quad \text{conditions on } g, g', R, R' .$$

To prove this theorem we first determine some stabilizers, which will be also useful at many other places. To see why they are important please look first at Proposition 2.17.

Definition 2.13. Let V be a subspace of $\bigoplus_{\Lambda \in P^+} L(\Lambda)$, and let Y be a submonoid of $\widehat{G}$. Set

$$\begin{aligned} \mathrm{N}_Y(V) &:= \{ x \in Y \mid xV = V \} , \\ \mathrm{M}_Y^{\subseteq}(V) &:= \{ x \in Y \mid xV \subseteq V \} , \\ \mathrm{M}_Y^{\supseteq}(V) &:= \{ x \in Y \mid xV \supseteq V \} , \\ \mathrm{Z}_Y(V) &:= \{ x \in Y \mid \forall v \in V : \ xv = v \} . \end{aligned}$$

Remarks 2.14. Let V be a subspace of $\bigoplus_{\Lambda \in P^+} L(\Lambda)$. Denote by $V^\perp$ the orthogonal complement of V. Let Y be a subgroup of G.

(1) If $\mathrm{M}_Y^{\subseteq}(V)$ or $\mathrm{M}_Y^{\supseteq}(V)$ is a group, then $\mathrm{M}_Y^{\subseteq}(V) = \mathrm{M}_Y^{\supseteq}(V) = \mathrm{N}_Y(V)$.

(2) If $V^{\perp\perp} = V$, then we have

$$N_Y(V^\perp) = N_{Y^*}(V)^* , \quad M_Y^{\subseteq}(V^\perp) = M_{Y^*}^{\subseteq}(V)^* , \quad M_Y^{\supseteq}(V^\perp) = M_{Y^*}^{\supseteq}(V)^* .$$

Because of $B(R)^\perp = K(R)$ and $K(R)^\perp = B(R)$, we can take $V = B(R), K(R)$.

For the next theorem recall that we denote a set of simple roots $\{ \alpha_j \mid j \in J \}$, $J \subseteq I$, also by J.

Theorem 2.15. *Let $R = \tau R(\Theta)$ be a face of the Tits cone and $\alpha \in \Delta_{re}$. Then*

$$N_{U_\alpha}(B(R)) = M_{U_\alpha}^{\subseteq}(B(R)) = M_{U_\alpha}^{\supseteq}(B(R)) = \begin{cases} U_\alpha & \text{if} \quad \alpha \in \tau\left(\Delta_{re}^+ \cup \mathcal{W}_\Theta\Theta \cup \mathcal{W}_{\Theta^\perp}\Theta^\perp\right) \\ \{1\} & \text{else} \end{cases},$$

$$Z_{U_\alpha}(B(R)) = \begin{cases} U_\alpha & \text{if} \quad \alpha \in \tau\Big(\left(\Delta_{re}^+ \setminus \mathcal{W}_{\Theta^\perp}\Theta^\perp\right) \cup \mathcal{W}_\Theta\Theta\Big) \\ \{1\} & \text{else} \end{cases}.$$

PROOF. If $x_\alpha \in \mathbf{g}_\alpha$ and $v_\mu \in L(\Lambda)_\mu \cap B(R)$, then

$$(\exp x_\alpha)\, v_\mu = \sum_{k \in \mathbb{N}_0,\ \mu + k\alpha \in P(\Lambda)} \frac{x_\alpha^k}{k!}\, v_\mu .$$

To prove the theorem we have to investigate the half-strings $(\mu + \mathbb{N}_0\alpha) \cap P(\Lambda)$ appearing in such sums.

• At first we examine a full string $(\mu + \mathbb{Z}\alpha) \cap P(\Lambda)$ with $\mu \in P(\Lambda) \cap R$: If $\alpha \in \text{span}\,(R) \cap P$ then clearly:

$$\alpha\text{-string through } \mu \subseteq R \cap P . \tag{2.2}$$

If $\alpha \notin \text{span}\,(R) \cap P$ then:

$$\alpha\text{-string through } \mu = \begin{cases} \{\mu,\ \mu - \alpha, \dots \mu - \mu(h_\alpha)\alpha\} & \text{if} \quad \mu(h_\alpha) > 0 \\ \{\mu\} & \text{if} \quad \mu(h_\alpha) = 0 \\ \{\mu,\ \mu + \alpha, \dots \mu - \mu(h_\alpha)\alpha\} & \text{if} \quad \mu(h_\alpha) < 0 \end{cases}. \tag{2.3}$$

As an example we prove the case $\mu(h_\alpha) = 0$, the other two cases are proved by similar arguments. Suppose there exists an integer $z \in \mathbb{Z}\backslash\{0\}$, such that $\mu + z\alpha \in X$. Then also $\sigma_\alpha(\mu + z\alpha) = \mu - z\alpha \in X$. Because R is a face of X, and

$$(\mu + z\alpha) + (\mu - z\alpha) = 2\mu \in R$$

we get $\mu \pm z\alpha \in R$. Therefore

$$(\mu + z\alpha) - (\mu - z\alpha) = 2z\alpha \in \text{span}\,(R) ,$$

contradicting $\alpha \notin \text{span}\,(R)$.

• Next we determine the real roots α for which all the α-half strings starting in elements of $R \cap P(\Lambda)$ satisfy certain properties for all $\Lambda \in P^+$.

(1) The following statements are equivalent:

  (i) $\forall_{\substack{\Lambda \in P^+ \\ \mu \in R \cap P(\Lambda)}} : \quad (\mu + \mathbb{N}_0\alpha) \cap P(\Lambda) \subseteq R \cap P$

  (ii) Either

  $\alpha \in \text{span}\,(R) \cap P$

  or

  $\alpha \notin \text{span}\,(R) \cap P \quad$ and $\quad \forall\, \mu \in R \cap P : \ \mu(h_\alpha) \geq 0$.

  (iii) $\alpha \in \tau\left(\Delta_{re}^+ \cup \mathcal{W}_\Theta\Theta \cup \mathcal{W}_{\Theta^\perp}\Theta^\perp\right)$.

(2) The following statements are equivalent:

  (i) $\forall_{\substack{\Lambda \in P^+ \\ \mu \in R \cap P(\Lambda)}} : \quad (\mu + \mathbb{N}_0\alpha) \cap P(\Lambda) = \{\mu\}$.

$$\text{(ii)} \quad \alpha \notin \operatorname{span}(R) \cap P \quad \text{and} \quad \forall\, \mu \in R \cap P \, : \, \mu(h_\alpha) \geq 0 \; .$$

$$\text{(iii)} \quad \alpha \in \tau\Big((\Delta_{re}^+ \setminus \mathcal{W}_{\Theta^\perp}\Theta^\perp) \cup \mathcal{W}_\Theta \Theta \Big) \; .$$

Taking into account $\bigcup_{\Lambda \in P^+} P(\Lambda) = X \cap P$, the equivalence of (1i) and (1ii) can be read off from (2.2) and (2.3).

If (2ii) holds, then also (1ii) holds, and therefore also (1i). To show (2i) suppose there exist $\Lambda \in P^+$, $\mu \in R \cap P(\Lambda)$, $n \in \mathbb{N}$, such that $\mu + n\alpha \in R \cap P$. Then we have $\alpha \in \operatorname{span}(R) \cap P$, which contradicts (2ii).

If (2i) holds, then also (1i) holds, and therefore also (1ii). To show (2ii) suppose $\alpha \in \operatorname{span}(R) \cap P$. Then we have for all $\mu \in R \cap P$:

$$\begin{aligned} \mu - \mu(h_\alpha)\alpha &= \sigma_\alpha \mu \;\in\; \operatorname{span}(R) \cap X \;=\; R \; , \\ \sigma_\alpha \mu + \mu(h_\alpha)\alpha &= \mu \;\in\; R \; . \end{aligned}$$

By using (2i) and $\bigcup_{\Lambda \in P^+} P(\Lambda) = X \cap P$, we conclude $-\mu(h_\alpha) \leq 0$, $\mu(h_\alpha) \leq 0$, and therefore $\mu(h_\alpha) = 0$. We have shown that σ_α fixes every element of R, and also every element of $\operatorname{span}(R)$, which contradicts $\alpha \in \operatorname{span}(R)$.

It is sufficient to show the equivalence of (1ii), (1iii), and the equivalence of (2ii), (2iii) only for $R = R(\Theta)$. For these equivalences it is sufficient to show the following statements:

(a) $\alpha \in \mathcal{W}_\Theta \Theta \quad \Longleftrightarrow \quad \forall\, \mu \in R(\Theta) \cap P \, : \, \mu(h_\alpha) = 0$.

(b) $\alpha \in \mathcal{W}_{\Theta^\perp} \Theta^\perp \quad \Longleftrightarrow \quad \alpha \in \operatorname{span}(R(\Theta)) \cap P$.

(c) $\alpha \in \Delta_{re}^+ \setminus (\mathcal{W}_\Theta \Theta \cup \mathcal{W}_{\Theta^\perp}\Theta^\perp) \quad \Longleftrightarrow$
$\alpha \notin \operatorname{span}(R(\Theta)) \cap P$ and $\forall\, \mu \in R(\Theta) \cap P \, : \, \mu(h_\alpha) \geq 0$ and
$\exists\, \mu \in R(\Theta) \cap P \, : \, \mu(h_\alpha) > 0$.

Statement (a) is valid due to $\mathcal{W}_\Theta = Z_\mathcal{W}(R(\Theta))$. To show (b) note that

$$\text{(2.4)} \qquad \mathcal{W}_\Theta \, , \, \mathcal{W}_{\Theta^\perp} \subseteq \mathcal{W}_{\Theta \cup \Theta^\perp} = N_\mathcal{W}(R(\Theta)) = N_\mathcal{W}(\operatorname{span}(R(\Theta))) \, ,$$

where the last equality follows because for $\sigma \in N_\mathcal{W}(\operatorname{span}(R(\Theta)))$ we have

$$\begin{aligned} \sigma R(\Theta) &= (\sigma R(\Theta) - \sigma R(\Theta)) \cap X = \sigma\,(R(\Theta) - R(\Theta)) \cap X \\ &= (R(\Theta) - R(\Theta)) \cap X = R(\Theta) \; . \end{aligned}$$

The direction '$\Rightarrow$' of (b) follows from $\Theta^\perp \subseteq \operatorname{span}(R(\Theta))$ and (2.4). To the reverse direction: If $\alpha \in \operatorname{span}(R(\Theta)) \cap P$, then σ_α leaves $\operatorname{span}(R(\Theta))$ invariant. Due to (2.4) we find $\alpha \in \mathcal{W}_{\Theta \cup \Theta^\perp}(\Theta \cup \Theta^\perp) = \mathcal{W}_\Theta \Theta \,\dot\cup\, \mathcal{W}_{\Theta^\perp}\Theta^\perp$.

Suppose there exist $\sigma \in \mathcal{W}_\Theta$, $i \in \Theta$, such that $\sigma\alpha_i \in \operatorname{span}(R(\Theta))$. Because of (2.4) we find $\alpha_i \in \operatorname{span}(R(\Theta))$, which contradicts $\alpha_i(h_i) = 2$.

Now we show the direction '$\Rightarrow$' of (c): Due to (b) we find $\alpha \notin \operatorname{span}(R(\Theta)) \cap P$. Because of (a) there exists an element $\mu \in R(\Theta) \cap P$ such that $\mu(h_\alpha) \neq 0$.

Note that due to [**K,P 3**], Lemma 2.1, we have

$$\Delta_{re}^+ \setminus \mathcal{W}_{\Theta \cup \Theta^\perp}(\Theta \cup \Theta^\perp) = \bigcap_{\eta \in \mathcal{W}_{\Theta \cup \Theta^\perp}} \eta \Delta_{re}^+ \; .$$

Therefore this is a $\mathcal{W}_{\Theta \cup \Theta^\perp}$-invariant set of positive roots. For every $\mu \in R(\Theta) \cap P$ there exist $\sigma \in \mathcal{W}_{\Theta \cup \Theta^\perp}$, $\tilde\mu \in \overline{F}_\Theta \cap P$, such that $\mu = \sigma\tilde\mu$ and we find:

$$\mu(h_\alpha) = \sigma\tilde\mu(h_\alpha) = \tilde\mu(h_{\sigma^{-1}\alpha}) \geq 0$$

It remains to show the direction '$\Leftarrow$' of (c): Because of (a) and (b) we find $\alpha \notin \mathcal{W}_\Theta\Theta \cup \mathcal{W}_{\Theta^\perp}\Theta^\perp$. Suppose $\alpha \in \Delta_{re}^- \setminus (\mathcal{W}_\Theta\Theta \cup \mathcal{W}_{\Theta^\perp}\Theta^\perp)$. Then $(-\alpha) \in \Delta_{re}^+ \setminus (\mathcal{W}_\Theta\Theta \cup \mathcal{W}_{\Theta^\perp}\Theta^\perp)$, and due to (c) '$\Rightarrow$' there exists an element $\mu \in R(\Theta) \cap P$, such that $\mu(h_{-\alpha}) > 0$. Inserting $h_{-\alpha} = -h_\alpha$ leads to a contradiction.

• Now we can proof the first statement of the theorem. Because of Remark 2.14, (1) it is sufficient to show

$$\mathrm{M}_{U_\alpha}^{\subseteq}(B(R)) = \begin{cases} U_\alpha & \text{if} \quad \alpha \in \tau\left(\Delta_{re}^+ \cup \mathcal{W}_\Theta\Theta \cup \mathcal{W}_{\Theta^\perp}\Theta^\perp\right) \\ \{1\} & \text{else} \end{cases} .$$

Let $\alpha \in \tau\left(\Delta_{re}^+ \cup \mathcal{W}_\Theta\Theta \cup \mathcal{W}_{\Theta^\perp}\Theta^\perp\right)$. We have to show the inclusion '$\supseteq$'. If $x_\alpha \in \mathbf{g}_\alpha$ and $v_\mu \in L(\Lambda)_\mu \subseteq B(R)$, then

$$(\exp x_\alpha)\, v_\mu = \sum_{k\in\mathbb{N}_0,\ \mu+k\alpha\in P(\Lambda)} \underbrace{\frac{x_\alpha^k}{k!}\, v_\mu}_{\in L(\Lambda)_{\mu+k\alpha}} .$$

Due to the equivalence of (1iii) and (1i) we have $P(\Lambda) \cap (\mu + \mathbb{N}_0\alpha) \subseteq R \cap P$. Therefore $(\exp x_\alpha)v_\mu \in B(R)$.

Let $\alpha \notin \tau\left(\Delta_{re}^+ \cup \mathcal{W}_\Theta\Theta \cup \mathcal{W}_{\Theta^\perp}\Theta^\perp\right)$. We have to show the inclusion '$\subseteq$'. Because of the equivalence of (1i) and (1iii) there exist $\Lambda \in P^+$, $\mu \in P(\Lambda) \cap R$, such that $P(\Lambda) \cap (\mu + \mathbb{N}_0\alpha) \not\subseteq P \cap R$.

Now $\mathbf{s}_\alpha = \mathbf{g}_\alpha \oplus [\mathbf{g}_\alpha, \mathbf{g}_{-\alpha}] \oplus \mathbf{g}_{-\alpha}$ is isomorphic to $sl(2,\mathbb{F})$, and there exists a decomposition of $L(\Lambda)$ in a direct sum of irreducible, finite dimensional $\mathbf{s}_\alpha$-modules, which are $\mathbf{h}$-invariant. Because the α-string through μ is finite, there is a $\mathbf{s}_\alpha$-module V among these modules, such that the α-string through μ restricted to $\mathbb{F}h_\alpha$ gives the set of weights of V.

Suppose there exists an element $x_\alpha \in \mathbf{g}_\alpha \setminus \{0\}$ such that $\exp(x_\alpha) \in \mathrm{M}_{U_\alpha}^{\subseteq}(B(R))$. Then for $v_\mu \in V \cap L(\Lambda)_\mu \setminus \{0\}$ we have

$$B(R) \ni (\exp x_\alpha)\, v_\mu = \sum_{k\in\mathbb{N}_0,\ \mu+k\alpha\in P(\Lambda)} \underbrace{\frac{x_\alpha^k}{k!}\, v_\mu}_{\in L(\Lambda)_{\mu+k\alpha}\setminus\{0\}} ,$$

which is a contradiction.

• The second statement of the theorem is proved similar to the first. Use the equivalence of (2i) and (2iii) instead of the equivalence of (1i) and (1iii).

□

Theorem 2.16. *Let $R = \tau R(\Theta)$ be a face of the Tits cone X. We have:*

(1a) $M_T^{\subseteq}(B(R)) = M_T^{\supseteq}(B(R)) = N_T(B(R)) = T$.

(1b) $Z_T(B(R)) = \tau T_\Theta \tau^{-1}$.

(2a) $M_N^{\subseteq}(B(R)) = M_N^{\supseteq}(B(R)) = N_N(B(R)) = \tau \tilde{N}_{\Theta\cup\Theta^\perp} \tau^{-1}$.

(2b) $Z_N(B(R)) = \tau N_\Theta \tau^{-1}$.

(3a) $M_G^{\subseteq}(B(R)) = M_G^{\supseteq}(B(R)) = N_G(B(R)) = \tau P_{\Theta\cup\Theta^\perp} \tau^{-1}$.

(3b) $Z_G(B(R)) = \tau\left(G_\Theta \ltimes U^{\Theta\cup\Theta^\perp}\right)\tau^{-1}$.

PROOF. Because of Remark 2.14, (1) we have to show (1a), (2a), and (3a) only for $\mathrm{M}_Y^{\subseteq}(B(R))$. Because of

$$\begin{aligned} \mathrm{M}_Y^{\subseteq}(B(\tau R(\Theta))) &= \mathrm{M}_Y^{\subseteq}(n_\tau B(R(\Theta))) = n_\tau\, \mathrm{M}_Y^{\subseteq}(B(R(\Theta)))\, n_\tau^{-1} , \\ \mathrm{Z}_Y(B(\tau R(\Theta))) &= \mathrm{Z}_Y(n_\tau B(R(\Theta))) = n_\tau\, \mathrm{Z}_Y(B(R(\Theta)))\, n_\tau^{-1} , \end{aligned}$$

we can restrict to the cases $R = R(\Theta)$, $Y = T,\ N,\ G$.

Obviously **(1a)** holds. **To (1b).** Write $t \in T$ in the form $t = \prod_{i=1}^{2n-l} t_i(s_i)$ where $s_i \in \mathbb{F}^\times$. Then $t \in \mathrm{Z}_T(B(R(\Theta)))$ if and only if $\prod_i s_i^{\lambda(h_i)} = 1$ for all $\lambda \in R(\Theta) \cap P$. Because of $R(\Theta) = \{\, \lambda \in X \mid \lambda(h_k) = 0\,,\ k \in \Theta \,\} \ni \Lambda_i$ for all $i \notin \Theta$, this is equivalent to $s_i = 1$ for all $i \notin \Theta$.

To (2a). Let $n_\sigma \in N$. Comparing the decompositions into weight spaces, we conclude that $n_\sigma B(R(\Theta)) \subseteq B(R(\Theta))$ is equivalent to $\sigma R(\Theta) \subseteq R(\Theta)$. Due to Proposition 2.5 this is equivalent to $\sigma \in \mathcal{W}_{\Theta\cup\Theta^\perp}$.

To (2b). N_Θ is generated by T_Θ and $n_i(1) = \exp(e_i)\exp(-f_i)\exp(e_i)$, $i \in \Theta$. Due to (1b) and Theorem 2.15 we get $\mathrm{Z}_N(B(R(\Theta))) \supseteq N_\Theta$. To show the reverse inclusion write $n_\sigma \in \mathrm{Z}_N(B(R(\Theta)))$ in the form

$$n_\sigma = t n_{i_1}(1)\cdots n_{i_k}(1) \quad \text{where} \quad t \in T\,,\ \sigma = \sigma_{i_1}\cdots\sigma_{i_k} \in \mathcal{W} \ \text{reduced} .$$

Because n_σ fixes the weight spaces of $B(R(\Theta))$, we have $\sigma \in Z_{\mathcal{W}}(R(\Theta)) = \mathcal{W}_\Theta$, from which we conclude $i_1, \ldots, i_k \in \Theta$, compare [**Hu**], Chapter 5.5.

Because $n_{i_1}(1), \ldots\ n_{i_k}(1)$, n_σ are elements of the group $\mathrm{Z}_N(B(R(\Theta)))$, we get $t \in \mathrm{Z}_N(B(R(\Theta))) \cap T = \mathrm{Z}_T(B(R(\Theta))) = T_\Theta$.

To (3a). U^+ is generated by U_α, $\alpha \in \Delta_{re}^+$. Using Theorem 2.15 we get

$$u\,B(R(\Theta)) = B(R(\Theta)) \quad \text{for all} \quad u \in U^+ . \tag{2.5}$$

Due to the Bruhat decomposition of G, an element $g \in G$ can be written in the form $g = un\tilde{u}$ with $u, \tilde{u} \in U^+$, $n \in N$. Using (2.5) and (2a) we find

$$g \in \mathrm{M}_G^{\subseteq}(B(R(\Theta))) \iff n\,\underbrace{(\tilde{u}B(R(\Theta)))}_{=B(R(\Theta))} \subseteq \underbrace{u^{-1}B(R(\Theta))}_{=B(R(\Theta))} \iff n \in \tilde{N}_{\Theta\cup\Theta^\perp} .$$

Therefore $\mathrm{M}_G^{\subseteq}(B(R(\Theta))) = U^+\tilde{N}_{\Theta\cup\Theta^\perp}U^+ = P_{\Theta\cup\Theta^\perp}$.

To (3b). G_Θ is generated by U_α, $\alpha \in \pm\Theta$. By using Theorem 2.15 we get

$$G_\Theta \subseteq Z_G(B(R(\Theta))) .$$

The group $U^{\Theta\cup\Theta^\perp}$ is the normal subgroup of $P_{\Theta\cup\Theta^\perp}$ generated by the root groups U_α, $\alpha \in \Delta_{re}^+ \setminus (\mathcal{W}_\Theta\Theta \cup \mathcal{W}_{\Theta^\perp}\Theta^\perp)$. Due to Theorem 2.15 these root groups are subgroups of the group $\mathrm{Z}_G(B(R(\Theta)))$, which is normal in $\mathrm{N}_G(B(R(\Theta))) = P_{\Theta\cup\Theta^\perp}$. Therefore we conclude

$$U^{\Theta\cup\Theta^\perp} \subseteq \mathrm{Z}_G(B(R(\Theta))) .$$

Now let $g \in \mathrm{Z}_G(B(R(\Theta)))$. Because of $\mathrm{Z}_G(B(R(\Theta))) \subseteq \mathrm{N}_G(B(R(\Theta))) = P_{\Theta\cup\Theta^\perp}$, and because of the decompositions

$$\begin{aligned} P_{\Theta\cup\Theta^\perp} &= \tilde{G}_{\Theta\cup\Theta^\perp} \ltimes U^{\Theta\cup\Theta^\perp} , \\ \tilde{G}_{\Theta\cup\Theta^\perp} &= (G_\Theta \times G_{\Theta^\perp}) \rtimes (T_{I\setminus(\Theta\cup\Theta^\perp)}T_{rest}) , \end{aligned}$$

there exist elements $x \in G_{\Theta^\perp}T_{I\setminus(\Theta\cup\Theta^\perp)}T_{rest}$, $y \in G_\Theta \ltimes U^{\Theta\cup\Theta^\perp}$ such that $g = xy$, and furthermore $x = gy^{-1} \in \mathrm{Z}_G(B(R(\Theta)))$.

We have to show $x = 1$. Because of $U^+ \cap U^- = \{1\}$, it is sufficient to show $x, x^* \in U_{\Theta^\perp}$.

(α) At first we show $x \in U_{\Theta^\perp}$: Because of the Bruhat decomposition of $G_{\Theta^\perp} T_{I\setminus(\Theta\cup\Theta^\perp)} T_{rest}$, we can write x as a product

$$x = u n_\sigma \tilde{u} \quad \text{where} \quad u,\, \tilde{u} \in U_{\Theta^\perp}\,, \quad n_\sigma \in N_{\Theta^\perp} T_{I\setminus(\Theta\cup\Theta^\perp)} T_{rest} \ .$$

By applying x on $v_\Lambda \in L(\Lambda)_\Lambda \setminus \{0\}$, $\Lambda \in \overline{F}_\Theta \cap P$, we get

$$v_\Lambda = u n_\sigma \tilde{u} v_\Lambda = \underbrace{n_\sigma v_\Lambda}_{\neq 0} + \sum_{q \in Q^+} v_{\sigma\Lambda+q} \ .$$

Comparing the components of the weight spaces we find

$$n_\sigma v_\Lambda = v_\Lambda \ . \tag{2.6}$$

Because we have $\sigma\Lambda = \Lambda$ for all $\Lambda \in \overline{F}_\Theta \cap P$, we conclude $\sigma \in \mathcal{W}_\Theta \cap \mathcal{W}_{\Theta^\perp} = \{1\}$. Write $n_\sigma = n_1 \in T_{I\setminus(\Theta\cup\Theta^\perp)} T_{rest} T_{\Theta^\perp}$ in the form

$$n_\sigma = \prod_{i \notin \Theta,\, i=1,\ldots 2n-l} t_i(s_i) \quad , \quad s_i \in \mathbb{F}^\times \ ,$$

and insert in (2.6). Since $\Lambda_k \in \overline{F}_\Theta \cap P$ for all $k \in \{1, \ldots, 2n-l\} \setminus \Theta$, we find

$$1 = s_k \quad \text{for all} \quad k \in \{1, \ldots, 2n-l\} \setminus \Theta \ .$$

Therfore we get $n_\sigma = 1$, and $x = u\tilde{u} \in U_{\Theta^\perp}$.

(β) To show $x^* \in U_{\Theta^\perp}$ it is sufficient to show $x^* \in \mathrm{Z}_G\left(B(R(\Theta))\right)$, because then we can apply (α) to x^*.

Let $v \in B(R(\Theta))$. Because of $x \in \mathrm{Z}_G\left(B(R(\Theta))\right)$, we find

$$\langle\langle x^* v - v \mid v' \rangle\rangle = \langle\langle v \mid x v' - v' \rangle\rangle = 0 \quad \text{for all} \quad v' \in B(R(\Theta)) \ .$$

We get $x^* v - v \in (B(R(\Theta)))^\perp = K(R(\Theta))$.

Because of (3a) we have $x^* \in (U_{\Theta^\perp})^* \subseteq \mathrm{N}_G\left(B(R(\Theta))\right)$, and therefore we also get $x^* v - v \in B(R(\Theta))$.

Because $B(R(\Theta))$ and $K(R(\Theta))$ intersect trivially, we conclude $x^* v = v$.

□

Proposition 2.17. *Let R, S be faces of the Tits cone. Let Y be a subgroup of G, and $y \in Y$. We have:*

$$ye(R) = e(S) \iff R = S \ \text{ and } \ y \in Z_Y\left(B(R)\right) \ . \tag{2.7}$$

$$e(R)y = e(S) \iff R = S \ \text{ and } \ y \in Z_Y\left(B(R)\right)^* \ . \tag{2.8}$$

$$ye(R)y^{-1} = e(R) \iff y \in N_Y\left(B(R)\right) \cap N_{Y^*}\left(B(R)\right)^* \ . \tag{2.9}$$

Remarks 2.18. (1) For $R = \tau R(\Theta)$ we get due to Theorem 2.16:

Y	T	N	G
$\mathrm{Z}_Y\left(B(R)\right)$	$\tau\, T_\Theta\, \tau^{-1}$	$\tau\, N_\Theta\, \tau^{-1}$	$\tau\left(G_\Theta \ltimes U^{\Theta\cup\Theta^\perp}\right)\tau^{-1}$
$\mathrm{N}_Y\left(B(R)\right) \cap \mathrm{N}_{Y^*}\left(B(R)\right)^*$	T	$\tau\, \tilde{N}_{\Theta\cup\Theta^\perp}\, \tau^{-1}$	$\tau\, \tilde{G}_{\Theta\cup\Theta^\perp}\, \tau^{-1}$

(2) For $R = S = R(\Theta)$ and $Y = G$ this has been claimed by D. Peterson.

PROOF. We first show (2.7), then (2.8) follows by application of the Chevalley involution $*$. Let $ye(R) = e(S)$. The kernels $K(R)$, $K(S)$ of the two linear maps are the same. Comparing the decompositions of $K(R)$ and $K(S)$ into weight spaces, we find $P \cap (X \setminus R) = P \cap (X \setminus S)$. Therefore $R = S$.

Now $ye(R) = e(R)$ is equivalent to $ye(R)v = e(R)v$ for all $v \in \bigoplus_{\Lambda \in P^+} L(\Lambda)$, which is in turn equivalent to $y \in \mathrm{Z}_Y\left(B(R)\right)$.

Next we show show (2.9). The linear projections $ye(R)y^{-1}$, $e(R)$ are equal if and only if their images $yB(R)$, $B(R)$, and their kernels $yK(R)$, $K(R)$ are equal. This is equivalent to $y \in \mathrm{N}_Y\left(B(R)\right) \cap \mathrm{N}_Y\left(K(R)\right)$, and by Remark 2.14, (2) we get $\mathrm{N}_Y\left(K(R)\right) = \mathrm{N}_{Y^*}\left(B(R)\right)^*$.

□

Let Θ be special. Because of the Levi decomposition

$$P_{\Theta \cup \Theta^\perp} = \tilde{G}_{\Theta \cup \Theta^\perp} \ltimes U^{\Theta \cup \Theta^\perp} ,$$

and the decomposition

$$\tilde{G}_{\Theta \cup \Theta^\perp} = (G_\Theta \times G_{\Theta^\perp}) \rtimes T^{co}_\Theta , \quad \text{where} \quad T^{co}_\Theta := T_{I \setminus (\Theta \cup \Theta^\perp)} T_{rest} ,$$

we get the following decompositions:

$$\begin{aligned} P_{\Theta \cup \Theta^\perp} &= (G_{\Theta^\perp} \rtimes T^{co}_\Theta) \ltimes \left(G_\Theta \ltimes U^{\Theta \cup \Theta^\perp}\right) , \\ (P_{\Theta \cup \Theta^\perp})^* &= \left(G_\Theta \ltimes U^{\Theta \cup \Theta^\perp}\right)^* \rtimes (G_{\Theta^\perp} \rtimes T^{co}_\Theta) . \end{aligned}$$

The projections belonging to these semidirect products are denoted by

$$\begin{aligned} p_\Theta : P_{\Theta \cup \Theta^\perp} &\to G_{\Theta^\perp} \rtimes T^{co}_\Theta , \\ p^*_\Theta : (P_{\Theta \cup \Theta^\perp})^* &\to G_{\Theta^\perp} \rtimes T^{co}_\Theta . \end{aligned}$$

It is easy to see, that we have $p^*_\Theta = * \circ p_\Theta \circ *$, and $p^*_\Theta(x) = p_\Theta(x)$ for all $x \in P_{\Theta \cup \Theta^\perp} \cap (P_{\Theta \cup \Theta^\perp})^* = \tilde{G}_{\Theta \cup \Theta^\perp}$.

Proposition 2.19. *If $x \in P_{\Theta \cup \Theta^\perp}$, if $y \in (P_{\Theta \cup \Theta^\perp})^*$ then*

$$\begin{aligned} x\, e(R(\Theta)) &= p_\Theta(x)\, e(R(\Theta)) = e(R(\Theta))\, p_\Theta(x) , \\ e(R(\Theta))\, y &= e(R(\Theta))\, p^*_\Theta(y) = p^*_\Theta(y)\, e(R(\Theta)) . \end{aligned}$$

In particular, we have

$$\begin{aligned} U^+ e(R(\Theta)) &= U^+_{\Theta^\perp} e(R(\Theta)) = e(R(\Theta)) U^+_{\Theta^\perp} , \\ e(R(\Theta)) U^- &= e(R(\Theta)) U^-_{\Theta^\perp} = U^-_{\Theta^\perp} e(R(\Theta)) . \end{aligned}$$

Remark. D. Peterson claimed $U^+ e(R(\Theta)) \subseteq e(R(\Theta)) U^+$, and $e(R(\Theta)) U^- \subseteq U^- e(R(\Theta))$.

PROOF. Write $x \in P_{\Theta \cup \Theta^\perp}$ in the form

$$x = p_\Theta(x) a \quad \text{with} \quad a \in G_\Theta \ltimes U^{\Theta \cup \Theta^\perp} , \; p_\Theta(x) \in G_{\Theta^\perp} T^{co}_\Theta .$$

Due to the equivalences (2.7) and (2.9) of Proposition 2.17, and Remark 2.18, (1) we get

$$x e(R(\Theta)) = p_\Theta(x) \underbrace{a e(R(\Theta))}_{= e(R(\Theta))} = \underbrace{p_\Theta(x) e(R(\Theta)) p_\Theta(x)^{-1}}_{= e(R(\Theta))} p_\Theta(x) .$$

Because of the decompositions $U^+ = U_{\Theta\cup\Theta^\perp} \ltimes U^{\Theta\cup\Theta^\perp}$ and $U_{\Theta\cup\Theta^\perp} = U_{\Theta^\perp} \times U_\Theta$, we have $p_\Theta(U^+) = U_{\Theta^\perp}$.

The remaining equations follow by application of the Chevalley involution $*$.

□

The last two propositions are special cases of the following theorem, which is the main theorem for computations in $\widehat{G}$. In the next section we will see that this theorem allows to reduce any computation in $\widehat{G}$ to computations in G and $\mathcal{R}(X)$.

Theorem 2.20. **(a)** *Let $x, y \in G$, and let Θ, Ξ be special. The following statements are equivalent:*

(i) $\quad xe(R(\Theta)) \;=\; e(R(\Xi))y$.

(ii) $\Theta = \Xi$ *and* $\quad x \in P_{\Theta\cup\Theta^\perp}$, $y \in (P_{\Theta\cup\Theta^\perp})^*$ *with* $p_\Theta(x) \;=\; p^*_\Theta(y)$.

(b) *Let R be a face of the Tits cone, and $n_\sigma \in N$. Then*

$$n_\sigma e(R) n_\sigma^{-1} \;=\; e(\sigma R) \, .$$

PROOF. Part (b) has already been shown, and the direction '$(ii) \Rightarrow (i)$' of (a) is an easy consequence of the last proposition. For the reverse direction we show at first $\Theta = \Xi$: Comparing the images of $xe(R(\Theta))$ and $e(R(\Xi))y$ we get

$$x\, B(R(\Theta)) \;=\; B(R(\Xi)) \, .$$

Write x in the form $x = un_\sigma\tilde{u}$ with $u, \tilde{u} \in U^+$, $n_\sigma \in N$, and insert in this equation. Since $U^+ \subseteq \mathrm{N}_G\left(B(R(\Theta))\right), \mathrm{N}_G\left(B(R(\Xi))\right)$ we find

$$n_\sigma B(R(\Theta)) \;=\; B(R(\Xi)) \, .$$

Comparing the decompositions into weight spaces, we get $\sigma R(\Theta) \cap P = R(\Xi) \cap P$. From this follows $\Theta = \Xi$.

Now we can show the remaining statements of (ii). Comparing the images and kernels of $xe(R(\Theta))$ and $e(R(\Theta))y$, we find

$$xB(R(\Theta)) \;=\; B(R(\Theta)) \quad , \quad K(R(\Theta)) \;=\; y^{-1}K(R(\Theta)) \, .$$

Due to Theorem 2.16 and Remark 2.14 (2), we get $x \in P_{\Theta\cup\Theta^\perp}$, $y \in (P_{\Theta\cup\Theta^\perp})^*$. Using the last proposition and Proposition 2.17 we find:

$$\begin{aligned} & p_\Theta(x)e(R(\Theta)) \;=\; xe(R(\Theta)) \;=\; e(R(\Theta))y \;=\; p^*_\Theta(y)e(R(\Theta)) \\ \Rightarrow \;\; & \underbrace{(p_\Theta(x))^{-1} p^*_\Theta(y)}_{\in T^{co}_\Theta G_{\Theta^\perp}} \in G_\Theta \ltimes U^{\Theta\cup\Theta^\perp} \quad \Rightarrow \quad (p_\Theta(x))^{-1} p^*_\Theta(y) \;=\; 1 \, . \end{aligned}$$

□

As an easy consequence of the last theorem we get:

Corollary 2.21. *Let R, S be faces of the Tits cone X. Let $x, y \in G$. Then:*

$$xe(R) \;=\; e(S)y \quad \Rightarrow \quad \exists\, \tau \in \mathcal{W} : \;\; S \;=\; \tau R \, .$$

2.4. The unit regularity of $\widehat{G}$

For a monoid M denote by $M^\times$ its unit group, and by $Idem(M)$ its set of idempotents. M is called *unit regular*, if $M = M^\times Idem(M) = Idem(M)\, M^\times$.

Theorem 2.22. **(a)** *The Kac-Moody group G is the unit group of $\widehat{G}$.*
(b) *The set of idempotents of $\widehat{G}$ is given by*

$$\begin{aligned} Idem(\widehat{G}) &= \left\{ \, ge(R)g^{-1} \;\middle|\; g \in G\,,\; R \in \mathcal{R}(X) \, \right\} \\ &= \left\{ \, ge(R(\Theta))g^{-1} \;\middle|\; g \in G\,,\; \Theta \text{ special } \right\} . \end{aligned}$$

(c) *We have $\widehat{G} = G\,E\,G = G\,E_{sp}\,G$. In particular $\widehat{G}$ is unit regular.*

Remark. P. Slodowy already guessed $\widehat{G} = GE_{Sp}G$, D. Peterson claimed $\widehat{G} = GE_{Sp}G$, and $\text{Idem}(\widehat{G}) = \left\{ \, ge(R(\Theta))g^{-1} \;\middle|\; g \in G\,,\; \Theta \text{ special} \right\}$.

PROOF. To prove (c) it is sufficient to show $\widehat{G} \subseteq GE_{sp}G$. Due to the definition of $\widehat{G}$, and the formula $e(\sigma R(\Theta)) = n_\sigma e(R(\Theta))n_\sigma^{-1}$, an element of $\widehat{G}$ can be written in the form

$$g_1 e(R(\Theta_1)) \cdots g_p e(R(\Theta_p)) \qquad \text{where} \qquad g_1, \ldots, g_p \in G\,,\;\; \Theta_1, \ldots, \Theta_p \;\text{ special} \;.$$

We can transform this element in an element of $GE_{sp}G$ by applying $(p-1)$ times the following step, which uses the Birkhoff decomposition of G, Proposition 2.19, and Theorem 2.20 (b):

$$\begin{aligned} e(R(\Theta))u^- n_\sigma u^+ e(R(\tilde\Theta)) \;&=\; p^*_\Theta(u^-)e(R(\Theta))n_\sigma e(R(\tilde\Theta))p_{\tilde\Theta}(u^+) \\ =\; p^*_\Theta(u^-)\, e\left(R(\Theta) \cap \sigma R(\tilde\Theta)\right) n_\sigma p_{\tilde\Theta}(u^+) \;&=\; p^*_\Theta(u^-) n_\tau e(R(\Xi)) n_\tau^{-1} n_\sigma p_{\tilde\Theta}(u^+) \\ \text{where} \quad u^- \in U^-\,,\; n_\sigma \in N\,,\; u^+ &\in U^+\,, \;\text{ and } \;\; \Theta,\, \tilde\Theta \text{ special} \;. \end{aligned}$$

Obviously we have $(\widehat{G})^\times \supseteq G$. To show the reverse inclusion let $ge(R)h$ be a unit. Because $(\widehat{G})^\times$ is a group containing G, we get

$$e(R) \;=\; g^{-1}\,(ge(R)h)\,h^{-1} \;\in\; (\widehat{G})^\times \;.$$

Because of $e(R)^2 = e(R)$ we conclude

$$e(R) \;=\; e(R)^2 e(R)^{-1} \;=\; e(R)e(R)^{-1} \;=\; 1 \;.$$

Therefore $ge(R)h = gh \in G$.

Obviously $\text{Idem}(\widehat{G}) \supseteq \left\{ \, ge(R(\Theta))g^{-1} \;\middle|\; g \in G\,,\; \Theta \text{ special} \right\}$. To show the reverse inclusion let $ge(R(\Theta))h$ be idempotent. Then we have

$$e(R(\Theta))\, h\, g\, e(R(\Theta)) \;=\; e(R(\Theta)) \;.$$

Using the Birkhoff decomposition of G, we can write hg in the form

$$hg \;=\; v n_\sigma u \qquad \text{with} \qquad v \in U^-\,,\; n_\sigma \in N\,,\; u \in U^+ \;.$$

To cut short the notation set $p := p_\Theta$, $p^* := p^*_\Theta$. Due to Proposition 2.19 and Theorem 2.20 (b), we find

$$\begin{aligned} e(R(\Theta)) \;&=\; e(R(\Theta))\, v\, n_\sigma\, u\, e(R(\Theta)) \;=\; p^*(v)e(R(\Theta))\, n_\sigma\, e(R(\Theta))p(u) \\ &=\; p^*(v)\, e\left(R(\Theta) \cap \sigma R(\Theta)\right) n_\sigma p(u) \;. \end{aligned} \tag{2.10}$$

Due to Corollary 2.21 the faces $R(\Theta)$ and $R(\Theta) \cap \sigma R(\Theta)$ are $\mathcal{W}$-conjugate. In particular they are of the same dimension. Because $R(\Theta) \cap \sigma R(\Theta)$ is contained in

$R(\Theta)$ these two faces coincide. We conclude $R(\Theta) \subseteq \sigma R(\Theta)$. Applying the same argument once more, we find $R(\Theta) = \sigma R(\Theta)$. Therefore $\sigma \in \mathcal{W}_{\Theta \cup \Theta^\perp}$. Inserting in (2.10) and using Theorem 2.20 (a), we get

$$1 \;=\; p(p^*(v))p^*(n_\sigma p(u)) \;=\; p^*(v)p^*(n_\sigma)p(u) \;.$$

Now we have

$$\begin{aligned} ge(R(\Theta))h &= (gu^{-1})ue(R(\Theta))vn_\sigma(gu^{-1})^{-1} \\ &= (gu^{-1})p(u)e(R(\Theta))p^*(v)p^*(n_\sigma)p(u)p(u)^{-1}(gu^{-1})^{-1} \\ &= (gu^{-1})e(R(\Theta))(gu^{-1})^{-1} \;. \end{aligned}$$

□

Remark. With part (c) of the last theorem we have reached the description of the elements of $\widehat{G}$ by expressions of the form (2.1) of short length. Using Theorem 2.20 we can decide if two such expressions give the same element. We have even reached more. Using the length reduction step given in the proof of (c), we are able to compute the product of two such expressions. But we have to work with the projections p_Θ, which can be difficult.

2.5. The Weyl monoid $\widehat{\mathcal{W}}$ and the monoids $\widehat{T}$, $\widehat{N}$

In this section we first introduce and investigate the Weyl monoid $\widehat{\mathcal{W}}$, which plays the same role for the monoid $\widehat{G}$ as the Weyl group $\mathcal{W}$ for the Kac-Moody group G. It has similar structural properties as the Renner monoid of a reductive algebraic group. Later we will see that it is really the analogue of a Renner monoid.

The Weyl group acts on the monoid $(\mathcal{R}(X), \cap)$. The semidirect product $\mathcal{R}(X) \rtimes \mathcal{W}$ consists of the set $\mathcal{R}(X) \times \mathcal{W}$ with the structure of a monoid given by

$$(R,\sigma)\cdot(S,\tau) \;:=\; (R \cap \sigma S, \sigma\tau) \;.$$

It is easy to see that we get a congruence relation on $\mathcal{R}(X) \rtimes \mathcal{W}$ by

$$(R,\sigma) \sim (R',\sigma') \quad :\Longleftrightarrow \quad R = R' \quad \text{and} \quad \sigma'\sigma^{-1} \in Z_{\mathcal{W}}(R) \;.$$

We denote the congruence class of (R,σ) by $\varepsilon\,(R)\sigma$.

Definition 2.23. We call the monoid $\widehat{\mathcal{W}} := (\mathcal{R}(X) \rtimes \mathcal{W})/\sim$ the Weyl monoid.

It is easy to check, that we get embeddings of monoids

$$\begin{array}{ccc} \mathcal{R}(X) & \to & \widehat{\mathcal{W}} \\ R & \mapsto & \varepsilon\,(R)1 \end{array} \quad , \quad \begin{array}{ccc} \mathcal{W} & \to & \widehat{\mathcal{W}} \\ \sigma & \mapsto & \varepsilon\,(X)\sigma \end{array} \;.$$

We denote the element $\varepsilon\,(R)1$ by $\varepsilon\,(R)$, and the image of $\mathcal{R}(X)$ by $\mathcal{E}$. We identify $\mathcal{W}$ with its image in $\widehat{\mathcal{W}}$.

Proposition 2.24. *$\widehat{\mathcal{W}}$ is a unit regular monoid, with unit group $\mathcal{W}$ and set of idempotents $\mathcal{E}$.*

PROOF. Obviously $\varepsilon\,(X)\sigma$ is a unit with inverse $\varepsilon\,(X)\sigma^{-1}$. Now let $\varepsilon\,(R)\sigma \in (\widehat{\mathcal{W}})^\times$. Then there exists an element $\varepsilon\,(S)\tau \in \widehat{\mathcal{W}}$, such that

$$\varepsilon\,(X) \;=\; \varepsilon\,(R)\sigma \cdot \varepsilon\,(S)\tau \;=\; \varepsilon\,(R \cap \sigma S)\sigma\tau \;.$$

Therefore we get $X = R \cap \sigma S$. Because X is the biggest element of $(\mathcal{R}(X), \subseteq)$, we conclude $R = X$.

Obviously $\varepsilon(R)$ is idempotent. Now let $\varepsilon(R)\sigma$ be idempotent. Because of

$$\varepsilon(R)\sigma \;=\; (\varepsilon(R)\sigma)(\varepsilon(R)\sigma) \;=\; \varepsilon(R\cap\sigma R)\sigma^2\,,$$

we find $\sigma^2\sigma^{-1}=\sigma\in Z_{\mathcal{W}}(R)$. From this follows $\varepsilon(R)\sigma=\varepsilon(R)$.

Now the unit regularity follows from the definition of $\widehat{\mathcal{W}}$.

□

A monoid M is called an *inverse monoid*, if for every element $m\in M$ there exists a unique element $m^{inv}\in M$, such that $m\,m^{inv}\,m=m$ and $m^{inv}\,m\,m^{inv}=m^{inv}$. The map ${}^{inv}: M\to M$ is an involution extending the inverse map of the unit group.

Proposition 2.25. *$\widehat{\mathcal{W}}$ is an inverse monoid.*

PROOF. It is easy to check that $\varepsilon\left(\sigma^{-1}R\right)\sigma^{-1}$ is an inverse of $\varepsilon(R)\sigma\in\widehat{\mathcal{W}}$. Furthermore the idempotents of $\widehat{\mathcal{W}}$ commute. Due to [**How**], Theorem 5.1.1, $\widehat{\mathcal{W}}$ is an inverse monoid.

□

The following two propositions are not difficult to prove:

Proposition 2.26. *For $J\subseteq I$ set $\mathcal{E}_J:=\{\,\varepsilon(R)\mid R\in\mathcal{R}(X)_J\,\}$. Then*

$$\widehat{\mathcal{W}}_J \;:=\; \mathcal{W}_J\,\mathcal{E}_J \;=\; \mathcal{E}_J\,\mathcal{W}_J$$

is a submonoid of $\widehat{\mathcal{W}}$. For J, $K\subseteq I$ we have $\widehat{\mathcal{W}}_J\subseteq\widehat{\mathcal{W}}_K$ if and only if $J\subseteq K$.

We call $\widehat{\mathcal{W}}_J$ a *standard parabolic submonoid* of $\widehat{\mathcal{W}}$, and every conjugate $\sigma\widehat{\mathcal{W}}_J\sigma^{-1}$, $\sigma\in\mathcal{W}$, a *parabolic submonoid* of $\widehat{\mathcal{W}}$.

Proposition 2.27. *$\widehat{\mathcal{W}}$ acts faithfully on the Tits cone by*

$$(\sigma\varepsilon(R))\,\lambda \;:=\; \begin{cases}\sigma\lambda & \text{if} \quad \lambda\in R\\ 0 & \text{if} \quad \lambda\in X\setminus R\end{cases}\,, \qquad \sigma\varepsilon(R)\,\in\,\widehat{\mathcal{W}}\,.$$

The parabolic submonoid $\widehat{\mathcal{W}}_J$ is the stabilizer of any element of the facet F_J, as well as the stabilizer of F_J as a whole, $J\subseteq I$.

Next we introduce the monoids $\widehat{T}$ and $\widehat{N}$. Later we will see, that $\widehat{T}$ can be described as a generalized affine toric variety.

Definition 2.28. Denote by $\widehat{T}$ the monoid generated by T and E, and by $\widehat{N}$ the monoid generated by N and E.

It is not difficult to prove the following proposition, describing the structure of $\widehat{T}$ and $\widehat{N}$:

Proposition 2.29. *$\widehat{T}$ is an abelian inverse unit regular monoid with unit group T and set of idempotents E. $\widehat{N}$ is an inverse unit regular monoid with unit group N and set of idempotents E.*

The Weyl group is isomorphic to N/T. There is a similar description for the Weyl monoid. We get a congruence relation on $\widehat{N}$ as follows:

$$\widehat{n}\sim\widehat{n}' \quad :\Longleftrightarrow \quad \widehat{n}T=\widehat{n}'T \quad \Longleftrightarrow \quad \widehat{n}'\in\widehat{n}T \quad \Longleftrightarrow \quad \widehat{n}\in\widehat{n}'T$$

Set $\widehat{N}/T:=\widehat{N}/\sim$. In a similar way define $\widehat{T}/T$.

Proposition 2.30. *The monoid $\widehat{N}/T$ is isomorphic to $\widehat{\mathcal{W}}$ by:*

$$\begin{array}{rccc} \kappa: & \widehat{\mathcal{W}} & \to & \widehat{N}/T \\ & \sigma\varepsilon(R) & \mapsto & n_\sigma e(R)T \end{array}$$

The set of idempotents $\mathcal{E}$ of $\widehat{\mathcal{W}}$ is mapped onto $\widehat{T}/T$.

Proof. To show that κ is well defined and injective let $R = \tau R(\Theta)$, and $R' \in \mathcal{R}(X)$. Because $\widehat{T}$ is abelian, because of equivalence (2.7) of Theorem 2.17, Remark 2.18, (1), and $Z_{\mathcal{W}}(R) = \tau\mathcal{W}_\Theta\tau^{-1}$, we find:

$$\begin{aligned} n_\sigma e(R)T = \tilde{n}_{\sigma'}e(R')T \quad &\iff \quad \exists\, t \in T : \ n_\sigma^{-1}\tilde{n}_{\sigma'}te(R') = e(R) \\ \iff \quad R = R' \ \text{ and } \ &\exists\, t \in T : \ n_\sigma^{-1}\tilde{n}_{\sigma'}t \in \tau N_\Theta \tau^{-1} \\ \iff \quad R = R' \ \text{ and } \ &\sigma^{-1}\sigma' \in \tau\mathcal{W}_\Theta\tau^{-1} \quad \iff \quad \sigma\varepsilon(R) = \sigma'\varepsilon(R') \end{aligned}$$

Due to the last proposition, κ is surjective. It is a homomorphism of monoids because of $\kappa(e(X)) = T$, and

$$\begin{aligned} \kappa(\sigma\varepsilon(R)\cdot\tau\varepsilon(S)) &= \kappa(\sigma\tau\,\varepsilon(\tau^{-1}R \cap S)) = n_{\sigma\tau}Te(\tau^{-1}R \cap S) \\ &= (n_\sigma e(R) n_\tau e(S))\,T = \kappa(\sigma\varepsilon(R))\cdot\kappa(\tau\varepsilon(S)) . \end{aligned}$$

□

2.6. Some double coset partitions of $\widehat{G}$

In this section we determine for some subgroups of G easy representative systems of the corresponding double coset partitions of $\widehat{G}$.

Proposition 2.31. *We have*

$$\widehat{G} = \dot{\bigcup_{\Theta \text{ special}}} Ge(R(\Theta))G .$$

Proof. This follows by combining Theorem 2.22 (c) and Theorem 2.20 (a). □

Proposition 2.32. *Let $\epsilon \in \{+,-\}$. We have*

$$\widehat{G} = \dot{\bigcup_{R \in \mathcal{R}(X)}} Ge(R)B^\epsilon = \dot{\bigcup_{R \in \mathcal{R}(X)}} B^\epsilon e(R)G .$$

Proof. Let Θ be special. By using the Bruhat and Birkhoff decompositions of G, Proposition 2.19, and Theorem 2.20 (b), we find

$$\begin{aligned} Ge(R(\Theta))G &= \bigcup_{\sigma \in \mathcal{W}} Ge(R(\Theta))B^- n_\sigma B^\epsilon = \bigcup_{\sigma \in \mathcal{W}} Ge(R(\Theta))n_\sigma B^\epsilon \\ &= \bigcup_{\sigma \in \mathcal{W}} Ge(\sigma^{-1}R(\Theta))B^\epsilon = \bigcup_{R \text{ of type } \Theta} Ge(R)B^\epsilon . \end{aligned}$$

To show that the last union is disjoint, let $Ge(\sigma R(\Theta))B^\epsilon = Ge(\tau R(\Theta))B^\epsilon$. Then there exist elements $g \in G$, $b \in B^\epsilon$ such that

$$g n_\sigma e(R(\Theta)) n_\sigma^{-1} b = n_\tau e(R(\Theta)) n_\tau^{-1} .$$

Using Theorem 2.20 (a), we conclude $bn_\tau \in n_\sigma(P_{\Theta\cup\Theta^\perp})^*$. Because of the generalized Birkhoff and Bruhat decompositions

$$G = \dot{\bigcup_{x \in \mathcal{W}/\mathcal{W}_{\Theta\cup\Theta^\perp}}} B^\epsilon x (P_{\Theta\cup\Theta^\perp})^* ,$$

we get $\tau \in \sigma\mathcal{W}_{\Theta\cup\Theta^\perp} = \sigma N_{\mathcal{W}}(R(\Theta))$. Therefore $\tau R(\Theta) = \sigma R(\Theta)$.

Taking into account the previous proposition we have proved the first equation. The second follows by application of the Chevalley involution $*$.

□

Theorem 2.33. *There are the Bruhat and Birkhoff decompositions*

$$\widehat{G} = \dot{\bigcup_{\widehat{w} \in \widehat{\mathcal{W}}}} B^\epsilon \widehat{w} B^\delta = \dot{\bigcup_{\widehat{n} \in \widehat{N}}} U^\epsilon \widehat{n} U^\delta ,$$

$$\textit{where} \quad (\epsilon, \delta) = \underbrace{(+,+)\,,\,(-,-)}_{Bruhat}\,,\,\underbrace{(+,-)\,,\,(-,+)}_{Birkhoff} .$$

Remark. D. Peterson claimed $\widehat{G} = U^\pm \widehat{N} U^\pm$.

Proof. Let Θ be special. We denote by $\mathcal{W}^{\Theta\cup\Theta^\perp}$ the minimal coset representatives of $\mathcal{W}/\mathcal{W}_{\Theta\cup\Theta^\perp}$. Due to [**Hu**], Section 5.12, and [**K**], Lemma 3.11, they can be characterized by $\mathcal{W}^{\Theta\cup\Theta^\perp} = \{\ \sigma \in \mathcal{W} \mid \sigma\alpha_i \in \Delta^+_{re} \text{ for all } i \in \Theta \cup \Theta^\perp\ \}$.

Because of $N_{\mathcal{W}}(R(\Theta)) = \mathcal{W}_{\Theta\cup\Theta^\perp}$, the faces of X, which are conjugated to $R(\Theta)$, are of the form $\sigma R(\Theta)$, $\sigma \in \mathcal{W}^{\Theta\cup\Theta^\perp}$. Due to the previous proposition we have

$$\widehat{G} = \dot{\bigcup_{\Theta\ sp.\,,\,\sigma \in \mathcal{W}^{\Theta\cup\Theta^\perp}}} G e(\sigma R(\Theta)) B^\delta .$$

Now we transform the term $Ge(\sigma R(\Theta))B^\delta$, using Proposition 2.17, Remark 2.18, (1), Proposition 2.19, Theorem 2.20 (b), the generalized Bruhat and Birkhoff decompositions for G, and the Bruhat and Birkhoff decompositions for $G_{\Theta^\perp}$:

$$\begin{aligned}
Ge(\sigma R(\Theta))B^\delta &= Ge(R(\Theta))n_\sigma^{-1}B^\delta = \bigcup_{\tau\in\mathcal{W}^{\Theta\cup\Theta^\perp}} B^\epsilon n_\tau P_{\Theta\cup\Theta^\perp} e(R(\Theta)) n_\sigma^{-1} B^\delta \\
&= \bigcup_{\tau\in\mathcal{W}^{\Theta\cup\Theta^\perp}} B^\epsilon n_\tau G_{\Theta^\perp} T^{co}_\Theta e(R(\Theta)) n_\sigma^{-1} B^\delta \\
&= \bigcup_{\tau\in\mathcal{W}^{\Theta\cup\Theta^\perp}} B^\epsilon n_\tau U^\epsilon_{\Theta^\perp} N_{\Theta^\perp} U^\delta_{\Theta^\perp} e(R(\Theta)) n_\sigma^{-1} B^\delta \\
&= \bigcup_{\tau\in\mathcal{W}^{\Theta\cup\Theta^\perp}} B^\epsilon \underbrace{n_\tau U^\epsilon_{\Theta^\perp} n_\tau^{-1}}_{\subseteq U^\epsilon} n_\tau N_{\Theta^\perp} e(R(\Theta)) n_\sigma^{-1} \underbrace{n_\sigma U^\delta_{\Theta^\perp} n_\sigma^{-1}}_{\subseteq U^\delta} B^\delta \\
&= U^\epsilon N e(R(\Theta)) n_\sigma^{-1} U^\delta = \bigcup_{\widehat{n} \in Ne(\sigma R(\Theta))} U^\epsilon \widehat{n} U^\delta .
\end{aligned}$$

Before we show that the last union is disjoint, note that the depth of $\lambda \in P(\Lambda)$ is defined as

$$d_\Lambda(\lambda) := ht(\Lambda - \lambda) \in \mathbb{N}_0 .$$

Let $U^\epsilon n_\tau e(\sigma R(\Theta))U^\delta = U^\epsilon \tilde{n}_\eta e(\sigma R(\Theta))U^\delta$. Then there exist elements $u \in U^\epsilon$, $\tilde{u} \in U^\delta$, such that

$$e(\sigma R(\Theta))\tilde{u}^{-1} \quad = \quad n_\tau^{-1} u \tilde{n}_\eta e(\sigma R(\Theta)) \ . \tag{2.11}$$

Due to Theorem 2.20 (a) we have $u\tilde{n}_\eta \in n_\tau(\sigma P_{\Theta\cup\Theta^\perp}\sigma^{-1})$. Taking into account the generalized Bruhat and Birkhoff decompositions

$$G \quad = \quad \dot{\bigcup_{x \,\in\, \mathcal{W}/\sigma\mathcal{W}_{\Theta\cup\Theta^\perp}\sigma^{-1}}} \quad U^\epsilon x \left(\sigma P_{\Theta\cup\Theta^\perp}\sigma^{-1}\right) \ ,$$

we conclude

$$\tau^{-1}\eta \ \in \ \sigma\mathcal{W}_{\Theta\cup\Theta^\perp}\sigma^{-1} \ .$$

Because of $\bigcup_{\Lambda\in P^+} L(\Lambda) = X \cap P$, we may choose an element $\Lambda \in P^+$, such that $P(\Lambda) \cap \sigma\mathcal{W}_{\Theta^\perp}F_\Theta \neq \emptyset$. Note that this set is invariant under $\tau^{-1}\eta$. Furthermore choose an element $\lambda \in (P(\Lambda) \cap \sigma\mathcal{W}_{\Theta^\perp}F_\Theta\,)$ such that

$$\begin{aligned} d_\Lambda(\lambda) \quad &= \quad \min\Big(d_\Lambda\left(P(\Lambda) \cap \sigma\mathcal{W}_{\Theta^\perp}F_\Theta\,\right)\Big) \quad \text{if} \quad \delta \ = \ + \ , \\ d_\Lambda(\tau^{-1}\eta\lambda) \quad &= \quad \min\Big(d_\Lambda\left(P(\Lambda) \cap \sigma\mathcal{W}_{\Theta^\perp}F_\Theta\,\right)\Big) \quad \text{if} \quad \delta \ = \ - \ . \end{aligned}$$

If we apply both sides of (2.11) on $v_\lambda \in L(\Lambda)_\lambda \setminus \{0\}$, we get

$$n_\tau^{-1}u\tilde{n}_\eta v_\lambda \quad = \quad e(\sigma R(\Theta))\tilde{u}^{-1}v_\lambda \ .$$

Comparing the components of the weight space $L(\Lambda)_{\tau^{-1}\eta\lambda}$, we conclude that there exists a $q \in Q_0^\epsilon$ such that

$$\tau^{-1}\eta\lambda \quad = \quad \lambda + q \ . \tag{2.12}$$

Because of the minimality of $d_\Lambda(\lambda)$ for $\delta = +$, and the minimality of $d_\Lambda(\tau^{-1}\eta\lambda)$ for $\delta = -$, we find $q = 0$. Inserting in (2.12) we conclude

$$\tau^{-1}\eta \ \in \ \sigma\mathcal{W}_\Theta\sigma^{-1} \ = \ Z_\mathcal{W}(\sigma R(\Theta)) \ . \tag{2.13}$$

Applying both sides of (2.11) on any $w_\mu \in B(\sigma R(\Theta))_\mu$, $\mu \in \sigma R(\Theta) \cap P$, we get

$$e(\sigma R(\Theta))\tilde{u}^{-1}w_\mu \quad = \quad n_\tau^{-1}u\tilde{n}_\eta w_\mu \ .$$

Comparing the components of the weight μ, taking into account (2.13), we find

$$w_\mu \quad = \quad n_\tau^{-1}\tilde{n}_\eta w_\mu \ .$$

Therefore $n_\tau^{-1}\tilde{n}_\eta \in Z_N(B(\sigma R(\Theta)))$. Due to Proposition 2.17 we have

$$n_\tau e(\sigma R(\Theta)) \quad = \quad \tilde{n}_\eta e(\sigma R(\Theta)) \ .$$

□

2.7. Constructing $\widehat{G}$ from the twin root datum

The definition of the monoid $\widehat{G}$ makes use of the admissible highest weight representations. In the following theorem we give a representation independent description of $\widehat{G}$. It uses only the twin root datum $(G, (U_\alpha)\alpha \in \Delta_{re})$, and the monoid $(\mathcal{R}(X), \cap)$ equipped with its Weyl group action. We identify the monoid $(\mathcal{R}(X), \cap)$ with the isomorphic monoid E.

Denote by $G \sqcup E$ the free product (= coproduct) of the monoids G and E. We identify G, E with their images under the canonical injections $G \hookrightarrow G \sqcup E$, $E \hookrightarrow G \sqcup E$.

Theorem 2.34. *Let $\sim$ be the congruence relation on $G \sqcup E$ generated by the following relations:*

(1) *For every $R \in \mathcal{R}(X)$, $n_\sigma \in N$:* $\quad n_\sigma e(R) n_\sigma^{-1} \sim e(\sigma R)$

(2) *For every special set $\Theta \subseteq I$, $x \in U_\alpha$, $\alpha \in \Delta_{re}$:*

(a) *If* $\alpha \in \mathcal{W}_{\Theta\cup\Theta^\perp}(\Theta \cup \Theta^\perp)$: $\quad xe(R(\Theta))x^{-1} \sim e(R(\Theta))$

(b) *If* $\alpha \in (\Delta_{re}^+ \setminus \mathcal{W}_{\Theta^\perp}\Theta^\perp) \cup \mathcal{W}_\Theta\Theta$: $\quad xe(R(\Theta)) \sim e(R(\Theta))$

(c) *If* $\alpha \in (\Delta_{re}^- \setminus \mathcal{W}_{\Theta^\perp}\Theta^\perp) \cup \mathcal{W}_\Theta\Theta$: $\quad e(R(\Theta))x \sim e(R(\Theta))$

Then $\widehat{G}$ is isomorphic to $(G \sqcup E)/\sim$.

PROOF. Let Θ be special. At first we show that the following relations hold:

(α) For every $\quad x \in \tilde{G}_{\Theta\cup\Theta^\perp}$: $\quad xe(R(\Theta))x^{-1} \sim e(R(\Theta))$

(β) For every $\quad x \in G_\Theta \ltimes U^{\Theta\cup\Theta^\perp}$: $\quad xe(R(\Theta)) \sim e(R(\Theta))$

(γ) For every $\quad x \in \left(G_\Theta \ltimes U^{\Theta\cup\Theta^\perp}\right)^*$: $\quad e(R(\Theta))x \sim e(R(\Theta))$

(δ) For every $x \in P_{\Theta\cup\Theta^\perp}$, $y \in (P_{\Theta\cup\Theta^\perp})^*$ with $p_\Theta(x) = p_\Theta^*(y)$:

$$xe(R(\Theta)) \sim e(R(\Theta))y$$

The relations (α) are valid, because of (1), (2a), and the definition of $\tilde{G}_{\Theta\cup\Theta^\perp}$. To show that the relations (β) are valid note that an element $x \in G_\Theta \ltimes U^{\Theta\cup\Theta^\perp}$ is a product of factors of the form

(I) u_α with $\alpha \in \mathcal{W}_\Theta\Theta$,

(II) $u_{\alpha_1}u_{\beta_1}\cdots u_{\alpha_m}u_{\beta_m}\cdot u_\gamma \cdot u_{\beta_m}^{-1}u_{\alpha_m}^{-1}\cdots u_{\beta_1}^{-1}u_{\alpha_1}^{-1}$ with $\alpha_1, \ldots, \alpha_m \in \Delta_{re}^+\backslash\mathcal{W}_{\Theta^\perp}\Theta^\perp$, $\beta_1, \ldots, \beta_m \in \Delta_{re}^+\cap\mathcal{W}_{\Theta^\perp}\Theta^\perp$, $\gamma \in \Delta_{re}^+\backslash(\mathcal{W}_\Theta\Theta\cup\mathcal{W}_{\Theta^\perp}\Theta^\perp)$,

where $u_\delta \in U_\delta$. To simplify the notation of the expressions (II), factors u_δ belonging to the same root δ, at different distance from the central factor u_γ, may be different.

For an expression of the form (I), we have according to (2b):

$$u_\alpha e(R(\Theta)) \sim e(R(\Theta)) \ .$$

For an expression of the form (II), we have according to (2a), (2b):

$$\begin{aligned} &u_{\alpha_1}u_{\beta_1}\cdots u_{\alpha_m}u_{\beta_m}\cdot u_\gamma \cdot u_{\beta_m}^{-1}u_{\alpha_m}^{-1}\cdots u_{\beta_1}^{-1}u_{\alpha_1}^{-1}\, e(R(\Theta)) \\ \sim \quad & e(R(\Theta))\, u_{\beta_1}\cdots u_{\beta_m}\cdot u_{\beta_m}^{-1}\cdots u_{\beta_1}^{-1} \quad = \quad e(R(\Theta)) \ . \end{aligned}$$

Therefore the relations (β) are valid. Similarly (γ) follows from (2a), (2c).

Using (α), (β), and (γ), the relations (δ) are proved in the same way as Proposition 2.19.

Now we can show that $(G \sqcup E)/\sim$ is isomorphic to $\widehat{G}$. For $x \in G \sqcup E$ denote by $[x]$ its equivalence class in $(G \sqcup E)/\sim$. Due to Theorem 2.15, Proposition 2.17, Theorem 2.20 (b), and the definition of $\widehat{G}$, there exists a surjective homomorphism of monoids

$$\phi : (G \sqcup E)/\sim \quad \to \quad \widehat{G}$$

with $\phi([x]) = x$ for all $x \in G$, $x \in E$.

ϕ is also injective. As in the proof of Theorem 2.22 (c), we can show

$$(G \sqcup E)/\sim \quad = \quad \{\ [g][e(R(\Theta))][h] \mid g, h \in G\,,\ \Theta \text{ special}\ \}$$

by using (δ) and (1). Now let

$$ge(R(\Theta))h \;=\; \tilde{g}e(R(\tilde{\Theta}))\tilde{h}\,.$$

Due to Theorem 2.20 (a) we get

$$\Theta \;=\; \tilde{\Theta} \;\text{ and }\; \begin{array}{c}\tilde{g}^{-1}g \in P_{\Theta\cup\Theta^\perp}\\ \tilde{h}h^{-1} \in (P_{\Theta\cup\Theta^\perp})^*\end{array} \quad\text{with}\quad p_\Theta(\tilde{g}^{-1}g) \;=\; p_\Theta^*(\tilde{h}h^{-1})\,.$$

With (δ) follows

$$[g][e(R(\Theta))][h] \;=\; [\tilde{g}][e(R(\tilde{\Theta}))][\tilde{h}]\,.$$

□

Remarks. (1) The theorem allows to decide if an (anti-)automorphism of G can be extended to an (anti-)automorphism of $\widehat{G}$, compare [**M 1**] for examples.

(2) The theorem indicates how to define the monoid $\widehat{G}$ for a group with a twin root datum, as defined in [**Ti 2**].

2.8. The action of $\widehat{G}$ on the admissible modules of $\mathcal{O}$

Recall that $\mathcal{O}_{adm}$ denotes the full subcategory of $\mathcal{O}$, whose objects consist of admissible $\mathbf{g}$-modules. Due to the complete reducibility theorem, compare [**K**], Corollary 10.7, or [**Mo,Pi**], Section 6.5, every module of $\mathcal{O}_{adm}$ is a sum of highest weight modules $L(\Lambda)$, $\Lambda \in P^+$.

Denote the corresponding category of G-modules also by $\mathcal{O}_{adm}$.

Proposition 2.35. **(1)** *Every G-module V of $\mathcal{O}_{adm}$ extends to a $\widehat{G}$-module, the action of the idempotent $e(R)$, $R \in \mathcal{R}(X)$, given by*

$$e(R)v_\lambda \;=\; \left\{\begin{array}{ll} v_\lambda & \lambda \in R \\ 0 & \lambda \in X\setminus R\end{array}\right. \;,\quad v_\lambda \in V_\lambda\,,\ \lambda \in P(V)\,.$$

This extension is compatible with sums, tensor products, and submodules.

(2) *Every G-homomorphism between two G-modules of $\mathcal{O}_{adm}$ is also a $\widehat{G}$-homomorphism between the corresponding $\widehat{G}$-modules.*

Proof. Due to its definition the monoid $\widehat{G}$ acts on $L(\Lambda)$, $\Lambda \in P^+$. Choose a decomposition of V into a direct sum of such modules to define an action of $\widehat{G}$ on V. This action is independent of the chosen decomposition and it is given in (1).

Let V, W be two G-modules of $\mathcal{O}_{adm}$. We have $P(V),\, P(W) \subseteq X\cap P$. If R is a face of X, we find for $v_\lambda \in V_\lambda$, $w_\mu \in W_\mu$:

$$\begin{aligned}(e(R)v_\lambda)\otimes(e(R)w_\mu) \;&=\; \left\{\begin{array}{ccl} v_\lambda\otimes w_\mu & if & \lambda\in R \text{ and } \mu \in R\\ 0 & else & \end{array}\right\}\\ &=\; \left\{\begin{array}{ccl} v_\lambda\otimes w_\mu & if & \lambda+\mu\in R\\ 0 & else & \end{array}\right\} \;=\; e(R)(v_\lambda\otimes w_\mu)\ .\end{aligned}$$

Therefore the extension is compatible with tensor products. Let $\phi : V \to W$ a G-homomorphism. For (2) it is sufficient to show $\phi\circ e(R) = e(R)\circ\phi$, which can easily be checked on the weight spaces of V.

□

Proposition 2.36. *The monoid $\widehat{G}$ acts faithfully on $\bigoplus_{i=1}^{2n-l} L(\Lambda_i)$.*

PROOF. Let $i \in \{n+1, \ldots, 2n-l\}$. The modules

$$(L(\Lambda_i), \pi_{\Lambda_i}) \qquad \text{and} \qquad (L(-\Lambda_i), \pi_{-\Lambda_i})$$

are one dimensional. By checking on the elements of G and E, which generate $\widehat{G}$, it is easy to see, that for every $x \in \widehat{G}$ we have

$$\pi_{\Lambda_i}(x) = c(x)\, id_{L(\Lambda_i)} \quad \text{where} \quad c(x) \in \mathbb{F}^\times ,$$

and furthermore

$$\pi_{-\Lambda_i}(x) = \frac{1}{c(x)}\, id_{L(-\Lambda_i)} .$$

For $i \in \{n+1, \ldots, 2n-l\}$ and $n_i \in \mathbb{N}$ we define $L(\Lambda_i)^{\otimes(-n_i)} := L(-\Lambda_i)^{\otimes n_i}$. For $\Lambda \in P^+$ we define $L(\Lambda)^{\otimes 0} := L(0)$.

Now let $x, x' \in \widehat{G}$ such that $\pi_{\Lambda_i}(x) = \pi_{\Lambda_i}(x')$ for all $i = 1, \ldots, 2n-l$. Any element Λ of P^+ is of the form

$$\Lambda = \sum_{i=1}^{2n-l} m_i \Lambda_i \quad \text{where} \quad m_1, \ldots, m_n \in \mathbb{N}_0 \text{ and } m_{n+1}, \ldots, m_{2n-l} \in \mathbb{Z} .$$

The module $\bigotimes_{i=1}^{2n-l} L(\Lambda_i)^{\otimes m_i}$ is an module of $\mathcal{O}_{adm}$. Its submodule generated by $\bigotimes_{i=1}^{2n-l} (L(\Lambda_i)_{\Lambda_i})^{\otimes m_i}$ is an irreducible highest weight module of highest weight Λ. Because of the last proposition we find $\pi_\Lambda(x) = \pi_\Lambda(x')$.

□

2.9. The submonoids $\widehat{G}_J$ $(J \subseteq I)$

Let $\emptyset \neq J \subseteq I$. Choose an optimal realization for the generalized Cartan submatrix A_J, consisting of saturated sublattices $H(A_J) \subseteq H$ and $P(A_J) \subseteq P$, such that

$$(h_j)_{j \in J} \subseteq H(A_J) \qquad \text{and} \qquad (\alpha_j)_{j \in J} \subseteq P(A_J) .$$

We have $P = P(A_J) \oplus H(A_J)^\perp$. Denote by $q_J : P \to P(A_J)$ the corresponding projection.

We have an embedding of groups

$$G(A_J) \to G . \tag{2.14}$$

Due to Proposition 2.7 we also have an embedding of lattices $\mathcal{R}(X_{A_J}) \to \mathcal{R}(X)$, which induces an embedding of monoids

$$E(A_J) \to E , \tag{2.15}$$

whose image we denote by E_J. Our aim is to show, that the embeddings (2.14), (2.15) induce an embedding of the monoid $\widehat{G(A_J)}$ into $\widehat{G}$, and to describe the part $\widehat{G}_J$ of the image, which is independent of the chosen sublattices $H(A_J)$ and $P(A_J)$. The results of this section will be important later.

The next proposition is well known.

Proposition 2.37. *Let $\Lambda \in P^+$ and regard the $\mathbf{g}$-module $L(\Lambda)$ as a $\mathbf{g}(A_J)$-module.*

(1) *Define an equivalence relation on $P(\Lambda)$ by*

$$\lambda \sim \mu \quad :\Longleftrightarrow \quad \lambda - \mu \in Q_J \,.$$

For every equivalence class C, the space $V_C := \bigoplus_{\lambda\in C} L(\Lambda)_\lambda$ is an admissible $\mathbf{g}(A_J)$-module of the category $\mathcal{O}(\mathbf{g}(A_J))$.

In particular the $\mathbf{g}(A_J)$-module $L(\Lambda)$ decomposes in a direct sum of admissible irreducible highest weight modules.

(2) *The $\mathbf{g}(A_J)$-module*

$$L_J(\Lambda) \;:=\; U(\boldsymbol{n}_J^-)L(\Lambda)_\Lambda$$

is an admissible irreducible highest weight module of highest weight $q_J(\Lambda)$. Set $P(L_J(\Lambda)) := \{\, \lambda \in P(\Lambda) \mid L_J(\Lambda) \cap L(\Lambda)_\lambda \neq \{0\} \,\}$. Then

$$L_J(\Lambda) \;=\; \bigoplus_{\lambda\in P(L_J(\Lambda))} L(\Lambda)_\lambda$$

is the weight space decomposition of $L_J(\Lambda)$, where $L(\Lambda)_\lambda$ is the weight space of weight $q_J(\lambda)$.

The module $L_J(\Lambda)$ is an isotypical component of $V_{[\Lambda]}$.

Definition 2.38. Let $\emptyset \neq J \subseteq I$.

Let $\widehat{G}_J$ be the monoid generated by G_J and E_J. Set $\widehat{G}_\emptyset := \{\, e(X) \,\}$.

Let $\widehat{N}_J$ be the monoid generated by N_J and E_J. Set $\widehat{N}_\emptyset := \{\, e(X) \,\}$.

Let $\widehat{T}_J$ be the monoid generated by T_J and E_J. Set $\widehat{T}_\emptyset := \{\, e(X) \,\}$.

Remarks 2.39. (1) Similarly to Proposition 2.30 the standard parabolic submonoid $\widehat{\mathcal{W}}_J$ is isomorphic to $\widehat{N}_J/T_J$.

(2) It is easy to show that we have

$$\widehat{G} \;=\; \widehat{G}_I \rtimes T_{rest} \quad , \quad \widehat{N} \;=\; \widehat{N}_I \rtimes T_{rest} \quad , \quad \widehat{T} \;=\; \widehat{T}_I \times T_{rest} \,.$$

Proposition 2.40. *Let $\emptyset \neq J \subseteq I$. The embeddings (2.14) and (2.15) induce an embedding of $\widehat{G(A_J)}$ into $\widehat{G}$. Restricted to $\widehat{G(A_J)}_J$, this embedding is independent of the chosen sublattices, its image given by $\widehat{G}_J$.*

PROOF. Due to Propositions 2.37 and 2.35, the monoid $\widehat{G(A_J)}$ acts on $L(\Lambda)$, $\Lambda \in P^+$. It is easy to check, that this action is compatible with the embeddings (2.14), (2.15). Therefore we get a morphism of monoids

$$\widehat{G(A_J)} \to \widehat{G} \,. \tag{2.16}$$

Note that the elements of $\widehat{G}_I$ act as identity on the modules $L(\Lambda_i)$, $i = n+1, \ldots, 2n-l$. Due to Proposition 2.36, the monoid $\widehat{G}_I$ acts faithfully on the sum $\bigoplus_{i\in I} L(\Lambda_i)$. Therefore $\widehat{G(A_J)}_J$ acts faithfully on $\bigoplus_{j\in J} L_J(\Lambda_j)$, and the restriction of the morphism (2.16) to $\widehat{G(A_J)}_J$ is injective. Obviously the image of $\widehat{G(A_J)}_J$ is given by $\widehat{G}_J$.

Because $T(A_J)$ acts faithfully on $\bigoplus_{\Lambda\in P^+} L(\Lambda)$, we also get an embedding of $T(A_J)$ into T. Denote by $T_{rest,\,J}$ the image of $T(A_J)_{rest}$.

The morphism (2.16) maps $\widehat{G(A_J)} = \widehat{G(A_J)}_J \rtimes T(A_J)_{rest}$ onto $\widehat{G}_J T_{rest,\,J}$. It is easy to check, that we have $\widehat{G}_J \rtimes T_{rest,\,J}$. Therefore the morphism is injective.

□

Corollary 2.41. *Let $\emptyset \neq J \subseteq I$. Then:*

(a) *The unit group and the set of idempotents of $\widehat{G}_J$ are given by*

$$(\widehat{G}_J)^\times \;=\; G_J \;,\quad Idem(\widehat{G}_J) \;=\; \{\; ge(R)g^{-1} \mid g \in G_J\,,\; R \in \mathcal{R}(X)_J \;\}\;.$$

We have $\widehat{G}_J \;=\; G_J E_J G_J$. In particular $\widehat{G}_J$ is unit regular.

(b) *There are the Bruhat- and Birkhoff decompositions*

$$\widehat{G}_J \;=\; \dot{\bigcup_{\widehat{w} \,\in\, \widehat{\mathcal{W}}_J}} \; B_J^\epsilon \,\widehat{w}\, B_J^\delta \qquad , \qquad \epsilon, \delta \in \{+, -\}\;.$$

Proof. The statements of the corollary can be proved easily for $\widehat{G}_I$, by using the corresponding statements for $\widehat{G}$, and Remark 2.39, (2). Now the statements for $\widehat{G}_J$ follow from the last proposition.

□

Proposition 2.42. *Let $\emptyset \neq J \subseteq I$. The monoid $\widehat{G}_J$ has a zero if and only if J is special. In this case $e(R(J))$ is the zero of $\widehat{G}_J$.*

Remark. By using $\widehat{G} = \widehat{G}_I \rtimes T_{rest}$, we conclude that $\widehat{G}$ has a zero if and only if A is nonsingular and has no component of finite type. In this case the zero is given by $e(c)$, where $c = \{0\}$ is the edge of the Tits cone.

Proof. Due to Proposition 2.40 it is sufficient to prove this proposition only for the monoid $\widehat{G}_I$.

Let I be special. Then the face $R(I)$ coincides with the edge c of the Tits cone. We show that $e(c)$ is the zero of $\widehat{G}_I$, by checking this for G_I and E: Due to Proposition 2.17 we have $G_I e(c) = e(c) = e(c) G_I$. Because of $R \cap c = c$, we have $e(R)e(c) = e(c) = e(c)e(R)$ for all $R \in \mathcal{R}(X)$.

Now suppose that z is the zero of $\widehat{G}_I$. Due to the last corollary, z is of the form $z = ge(R(\Theta))h$, with $g, h \in G_I$ and Θ special. Because z is the zero we get $z = g^{-1}zh^{-1} = e(R(\Theta))$. We also have $T_I z = z$. Due to Proposition 2.17 we get $T_I \subseteq T_\Theta$, from which follows $\Theta = I$. In particular, I is special.

□

Now we can give the stabilizers of the highest weight spaces of the admissible highest weight modules, which we use later several times.

Proposition 2.43. *Let $J \subseteq I$. For $\Lambda \in P^+ \cap F_J$ we have:*

(a) $N_{\widehat{G}}(L(\Lambda)_\Lambda) \;=\; B\widehat{\mathcal{W}}_J B\;.$

(b) $Z_{\widehat{G}}(L(\Lambda)_\Lambda) \;=\; U^J \widehat{G}_J X_\Lambda U^J \;=\; U\widehat{N}_J X_\Lambda U$ *, where*

$$X_\Lambda \;:=\; \left\{\; \prod_{i=1,\,\ldots,\,2n-l,\, i \notin J} t_i(s_i) \in T_{I \setminus J} T_{rest} \;\middle|\; \prod_{i=1,\,\ldots,\,2n-l,\, i \notin J} s_i^{\Lambda(h_i)} \;=\; 1 \;\right\}.$$

Proof. The proofs of (a) and (b) are similar, using

$$N_G(L(\Lambda)_\Lambda) \;=\; B\mathcal{W}_J B = U\tilde{N}_J U \quad \text{and} \quad Z_G(L(\Lambda)_\Lambda) \;=\; UN_J X_\Lambda U\;.$$

As an example we prove (a): Due to the Bruhat decomposition of $\widehat{G}$ an element $\widehat{g} \in \widehat{G}$ can be written in the form $\widehat{g} = un_\sigma e(R)\tilde{u}$, where $u, \tilde{u} \in U$, $n_\sigma \in N$, and $R \in \mathcal{R}(X)$. By using $\Lambda \in F_{J^\infty \cup J^0} \subseteq ri(R(J^\infty))$ we find

$$\begin{aligned} \widehat{g}\,(L(\Lambda)_\Lambda) \;=\; L(\Lambda)_\Lambda \iff n_\sigma e(R) L(\Lambda)_\Lambda \;=\; L(\Lambda)_\Lambda \iff \\ \Lambda \in R \quad \text{and} \quad n_\sigma L(\Lambda)_\Lambda \;=\; L(\Lambda)_\Lambda \iff R \supseteq R(J^\infty) \quad \text{and} \quad n_\sigma \in \tilde{N}_J \;. \end{aligned}$$

□

For $\Lambda \in P^+$ set $L_\emptyset(\Lambda) := L(\Lambda)_\Lambda$. Later we also need the following proposition.

Proposition 2.44. *Let $J \subseteq I$ and $\Lambda \in P^+$. Then U^J fixes $L_J(\Lambda)$ pointwise.*

PROOF. The case $J = \emptyset$ is obvious, let $J \neq \emptyset$. Choose $v_\Lambda \in L(\Lambda)_\Lambda \setminus \{0\}$. Due to the definition of $L_J(\Lambda)$, it is sufficient to show the following statements:

S_0 : $\quad u\,v_\Lambda \;=\; v_\Lambda \quad \text{for all} \quad u \in U^J$.

S_p, $p \in \mathbb{N}$: $\quad u\,x_1 \cdots x_p v_\Lambda \;=\; x_1 \cdots x_p v_\Lambda \quad \text{for all} \quad u \in U^J,\ x_1, \ldots, x_p \in \mathbf{n}_J^-$.

The statement S_0 is valid. Now the induction step from S_p to S_{n+1}, $p \in \mathbb{N}_0$: Fix an element $u \in U^J$. Let $x_1, \ldots, x_p \in \mathbf{n}_J^-$. Let $m \in \mathbb{N}$ and $y_i \in \mathbf{g}_{\beta_i}$, $\beta_i \in (\Delta_J)_{re}^-$, $i = 1, \ldots, m$. For $t_1, \ldots, t_m \in \mathbb{F}$ set

$$u(t_1, \ldots, t_m) \;:=\; \exp(t_1 y_1) \cdots \exp(t_m y_m) \;\in\; P_J \;.$$

Because U^J is normal in P_J, we have $u(t_1, \ldots, t_m)^{-1}\, u\, u(t_1, \ldots, t_m) \in U^J$. Using the induction assumption, we find

$$\begin{aligned} & u\, u(t_1, \ldots, t_m)\, x_1 \cdots x_p v_\Lambda \\ = \;& u(t_1, \ldots, t_m)\, u(t_1, \ldots, t_m)^{-1}\, u\, u(t_1, \ldots, t_m)\, x_1 \cdots x_p v_\Lambda \\ = \;& u(t_1, \ldots, t_m)\, x_1 \cdots x_p v_\Lambda \;. \end{aligned}$$

Because the $y_1, \ldots, y_m$ act locally nilpotent, the left and right side of this equation is polynomial in $t_1, \ldots, t_m$. Because of $|\mathbb{F}| = \infty$ the coefficients of the monomial $t_1 \cdot t_2 \ldots \cdot t_m$ on the left and on the right are equal. We get

$$u\, y_1 \cdots y_m x_1 \cdots x_p v_\Lambda \;=\; y_1 \cdots y_m x_1 \cdots x_p v_\Lambda \;.$$

This equation is also valid, if we permute $y_1, \ldots, y_m$. We conclude

$$\begin{aligned} & u\,[y_1[y_2 \cdots [y_{m-1}, y_m] \cdots]]\, x_1 \cdots x_p v_\Lambda \\ = \;& [y_1[y_2 \cdots [y_{m-1}, y_m] \cdots]]\, x_1 \cdots x_p v_\Lambda \;. \end{aligned}$$

The Lie algebra $\mathbf{n}_J^-$ is generated by $\mathbf{g}_\beta$, $\beta \in (\Delta_J)_{re}^-$. Therefore we get

$$u\, y x_1 \cdots x_p v_\Lambda \;=\; y x_1 \cdots x_p v_\Lambda \qquad \text{for} \qquad y, x_1, \ldots, x_p \in \mathbf{n}_J^- \;.$$

□

2.10. The monoid $\widehat{G}$ for a decomposable matrix A

Let the generalized Cartan matrix A be decomposable, $A = A_{I_1} \oplus A_{I_2}$. Choose optimal realizations

$$\begin{aligned} H_1 \supseteq \{\, h_i \mid i \in I_1 \,\} \quad &, \quad P_1 \supseteq \{\, \alpha_i \mid i \in I_1 \,\} \;, \\ H_2 \supseteq \{\, h_i \mid i \in I_2 \,\} \quad &, \quad P_2 \supseteq \{\, \alpha_i \mid i \in I_2 \,\} \end{aligned}$$

of A_{I_1} and A_{I_2} as described in Section 1.1. Recall that $H = H_1 \oplus H_2$ and $P = P_1 \oplus P_2$.

Due to Proposition 2.40 we get embeddings of $\widehat{G(A_{I_1})}$ and $\widehat{G(A_{I_2})}$ into $\widehat{G}$. Denote by $\widehat{G}_1$ and $\widehat{G}_2$ its images.

Proposition 2.45. *We have $\widehat{G} = \widehat{G}_1 \times \widehat{G}_2$.*

PROOF. The Kac-Moody algebras $\mathbf{g}(A_{I_1})$ and $\mathbf{g}(A_{I_2})$ embed into $\mathbf{g}$. To simplify the notation we identify these Kac-Moody algebras with its images. Similarly we identify $\widehat{G(A_{I_1})}$ and $\widehat{G(A_{I_2})}$ with G_1 and G_2.

If $L_i(\Lambda_i)$ is an irreducible highest weight module of $\mathbf{g}(A_{I_i})$ of highest weight $\Lambda_i \in P_i$, $i = 1, 2$, then

$$L_1(\Lambda_1) \otimes L_2(\Lambda_2) \tag{2.17}$$

is an irreducible highest weight module of $\mathbf{g} = \mathbf{g}(A_{I_1}) \oplus \mathbf{g}(A_{I_2})$ of highest weight $\Lambda_1 + \Lambda_2$. Because of $P^+ = P_1^+ \oplus P_2^+$, all admissible irreducible highest weight modules of $\mathbf{g}$ are obtained in this way.

We have $G = G(A_{I_1})G(A_{I_2})$, and due to Proposition 2.8 we have also $E = E(A_{I_1})E(A_{I_2})$. Now let $g_1 \in G(A_{I_1})$, $g_2 \in G(A_{I_2})$, and $e_1 \in E(A_{I_1})$, $e_2 \in E(A_{I_2})$. The elements g_1g_2 and e_1e_2 act on a decomposable element $v_1 \otimes v_2$ of the module (2.17) by

$$g_1g_2(v_1 \otimes v_2) = g_1v_1 \otimes g_2v_2 \quad , \quad e_1e_2(v_1 \otimes v_2) = e_1v_1 \otimes e_2v_2 .$$

From this follows easily, that the elements of $\widehat{G(A_{I_1})}$ commute with the elements of $\widehat{G(A_{I_2})}$ in $\widehat{G}$, and $\widehat{G} = \widehat{G(A_{I_1})}\widehat{G(A_{I_2})}$.

Now let $\widehat{g}_1, \widehat{g}_1' \in \widehat{G(A_{I_1})}$ and $\widehat{g}_2, \widehat{g}_2' \in \widehat{G(A_{I_2})}$, such that $\widehat{g}_1\widehat{g}_2 = \widehat{g}_1'\widehat{g}_2'$. Regarding this equation on the modules $L_1(\Lambda_1) \otimes L_2(0)$, $\Lambda_1 \in P_1^+$, we find $\widehat{g}_1 = \widehat{g}_1'$. Similarly we get $\widehat{g}_2 = \widehat{g}_2'$.

□

CHAPTER 3

An algebraic geometric setting

In this chapter we develop an easy, infinite dimensional algebraic geometry analogous to elementary algebraic geometry, generalizing the topologized coordinate rings of $L(\Lambda)$, and of the Kostant cones $G(L(\Lambda)_\Lambda)$, $\Lambda \in P^+$, given in [**K,P 2**], Section 3A.

This algebraic geometric setting is useful, because the monoids, coordinate rings, and Lie algebras, which are obtained by generalized Tannaka-Krein reconstructions, fit into this context in a natural way. For an example you may look at the next chapter.

There is a technical point. In the next chapter, we characterize the monoid, which is obtained by a generalized Tannaka-Krein reconstruction from the category $\mathcal{O}_{adm}$ and its category of restricted duals, as a Zariski closure $\overline{G}$ in a very big space. We have $\widehat{G} \subseteq \overline{G}$, and one of our main aims in Chapter 5 is to show equality. This is a long way, and we have to work with $\widehat{G}$ before we really know, that $\widehat{G}$ is closed. For this reason we have to formulate our algebraic geometric concepts also for non-closed varieties.

Most proofs in this chapter are easy, or parallel to elementary algebraic geometry. They are omitted except the proofs of some propositions, which will be used later in central position. Some more details can be found in [**M 1**].

Recall that $\mathbb{K}$ denotes a field with $|\mathbb{K}| = \infty$. In this chapter all linear spaces are $\mathbb{K}$-linear spaces, all algebras are $\mathbb{K}$-algebras. For all topological statements in which $\mathbb{K}$ is involved, we assume $\mathbb{K}$ to be equipped with the discrete topology.

3.1. Varieties and pnc-varieties

The category of (pnc-)varieties: The category of (pnc-)varieties will be constructed as a full subcategory of the category of sets with coordinate rings topologized by filter of ideals, which we introduce first:

An object $(A, \mathbb{K}[A], \mathcal{F}_A)$ of this category consists of a point separating algebra $\mathbb{K}[A]$ of $\mathbb{K}$-valued functions on a non-empty set A. We call $\mathbb{K}[A]$ the coordinate ring of A. Furthermore, a filter $\mathcal{F}_A$ in the lattice of ideals of $\mathbb{K}[A]$, which we call a filter of ideals for short. The coordinate ring $\mathbb{K}[A]$ is a topological algebra with $\mathcal{F}_A$ as a basis of neighborhoods of the zero. (Recall that $\mathbb{K}$ is equipped with the discrete topology.) The set of open ideals coincides with $\mathcal{F}_A$.

A morphism of two such objects $(A, \mathbb{K}[A], \mathcal{F}_A)$ and $(B, \mathbb{K}[B], \mathcal{F}_B)$ is a map $\phi : A \to B$, such that the comorphism $\phi^* : \mathbb{K}[B] \to \mathbb{K}[A]$ exists and satisfies $(\phi^*)^{-1}(\mathcal{F}_A) \subseteq \mathcal{F}_B$, i.e., ϕ^* is continuous.

In our notation we emphasize the filter of ideals more than the topology, because some of the following constructions involve rather operations with the ideals of the filter than with the sets of the topology.

The varieties of our algebraic geometry are constructed as follows: Let V be a vector space, V^* its dual, and $X \subseteq V^*$ a subspace, which separates the points of V. The algebra generated by X gives a coordinate ring $\mathbb{K}_X[V]$ on V, which is isomorphic to the symmetric algebra in X. The filter of ideals $\mathcal{F}_V$ is given by the filterbase of ideals

$$\{\ I(U) \mid U \text{ a finite dimensional subspace of } V.\ \}\ ,$$

where $I(U)$ denotes the ideal of functions of $\mathbb{K}_X[V]$ vanishing on U.

A set $\emptyset \neq A \subseteq V$ is equipped with a coordinate ring together with a filter of ideals by restriction:

$$\mathbb{K}[A] := \mathbb{K}_X[V]|_A \quad , \quad \mathcal{F}_A := \{I|_A \mid I \in \mathcal{F}_V\}\ .$$

Denote by $\operatorname{Specm}\mathbb{K}[A]$ the set of 1-codimensional ideals of $\mathbb{K}[A]$, and define

$$F\text{-}\operatorname{Specm}\mathbb{K}[A] \quad := \quad \mathcal{F}_A \cap \operatorname{Specm}\mathbb{K}[A]\ .$$

These ideals are the kernels of the continuous homomorphisms of algebras from $\mathbb{K}[A]$ to $\mathbb{K}$. We have a bijective map

$$\begin{array}{ccc} \overline{A} & \rightarrow & F\text{-}\operatorname{Specm}\mathbb{K}[A] \\ a & \mapsto & I_{\overline{A}}(a)|_A \end{array} \ ,$$

where $\overline{A}$ is the Zariski closure of A, and $I_{\overline{A}}(a)$ the ideal of functions of $\mathbb{K}[\overline{A}]$, which vanish in a. In particular the map

$$\begin{array}{ccc} A & \rightarrow & F\text{-}\operatorname{Specm}\mathbb{K}[A] \\ a & \mapsto & I_A(a) \end{array}$$

is bijective if and only if A is Zariski closed in V. Due to this fact, the following definitions are meaningful:

Definition 3.1. A set with coordinate ring topologized by a filter of ideals isomorphic to such an object $(A, \mathbb{K}[A], \mathcal{F}_A)$ is called a pnc-variety. It is called a variety if A is Zariski closed. ('pnc' means 'possibly not closed'.)

Remarks 3.2. (1) Due to the filter of ideals every pnc-variety can be completed to a variety.

(2) If V is of infinite dimension there are many possibilities to choose a subspace X of V^*, which separates the points of V, and therefore there are many possible coordinate rings of V. This flexibility will be important later.

The situation is different if we restrict to a finite dimensional subspace U of V. Because $X|_U$ separates the points of U, there is only the possibility $X|_U = U^*$, and $\mathbb{K}[U]$ is the classical coordinate ring of U. The topology on $\mathbb{K}[U]$ is discrete because of $I(U)|_U = \{0\} \in \mathcal{F}_U$. A map between two discrete topological spaces is always continuous. Therefore the category of varieties includes the category of the classical affine algebraic varieties.

Tangent spaces and tangent maps: Define the *tangent space* of a (pnc)-variety $(A, \mathbb{K}[A], \mathcal{F}_A)$ at a point $a \in A$ by

$$T_a A \quad := \quad \{\, \delta \in \mathrm{Der}_a \mathbb{K}[A] \mid \exists I \in \mathcal{F}_A : \ I \subseteq \ \mathrm{kernel}\, \delta \,\} \quad .$$

These are the derivations in a which are continuous.

If $(A, \mathbb{K}[A], \mathcal{F}_A)$, $(B, \mathbb{K}[B], \mathcal{F}_B)$ are pnc-varieties, $a \in A$, and $\phi : A \to B$ is a morphism, we get a linear map, the *tangent map* at a, by:

$$\begin{array}{rccc} T_a\phi : & T_a A & \to & T_{\phi(a)} B \\ & \delta & \mapsto & \delta \circ \phi^* \end{array}$$

Example 3.3. Let V be a vector space, and $(V, \mathbb{K}_X[V], \mathcal{F}_V)$ the variety given by a subspace $X \subseteq V^*$, which separates the points of V. The tangent space $T_a V$ at $a \in V$ can be identified with V by means of the linear bijective map $V \to T_a V$, which assigns to an element $v \in V$ the derivation $\delta_v \in T_a V$ defined by

$$f(a + tv) \;=\; f(a) + t\delta_v(f) + O(t^2) \qquad , \qquad t \in \mathbb{K}\,, \quad f \in \mathbb{K}_X[V]\,.$$

In this way, the tangent space $T_a V$ is independent of the chosen point separating subspace $X \subseteq V^*$.

Various constructions: It is not difficult to see, that the following constructions do not leave the category of (pnc-)varieties:

1. Substructures: Let $(B, \mathbb{K}[B], \mathcal{F}_B)$ be a pnc-variety and $a \in A \subseteq B$. The pnc-variety $(A, \mathbb{K}[A], \mathcal{F}_A)$ is defined in the obvious way by restricting the functions of $\mathbb{K}[B]$ onto A. If B is a variety and A is Zariski closed in B, then A is also a variety.

The inclusion map $\iota : A \to B$ is a morphism, and its tangent map $T_a\iota : T_a A \to T_a B$ is injective with the image

$$T_a\iota(T_a A) \;=\; \{\ \delta \in T_a B \mid \ \ \mathrm{kernel}\, \delta \supseteq I_B(A)\ \} \ .$$

(Here $I_B(A)$ denotes the vanishing ideal of A in $\mathbb{K}[B]$.)

2. Products: Let $(C, \mathbb{K}[C], \mathcal{F}_C)$, $(D, \mathbb{K}[D], \mathcal{F}_D)$ be (pnc-)varieties and $c \in C$, $d \in D$. The algebra $\mathbb{K}[C] \otimes \mathbb{K}[D]$ is topologized by the filterbase of ideals

$$\{\ (I_C \otimes 1\,,\ 1 \otimes I_D) \ \mid \ I_C \in \mathcal{F}_C\,,\ I_D \in \mathcal{F}_D\ \}$$

(the bracket ( ) stands for 'the ideal generated by'). This gives in the obvious way a (pnc-)variety $(C \times D, \mathbb{K}[C \times D], \mathcal{F}_{C\times D})$ on $C \times D$, which is, together with the projections pr_C, pr_D, the product of $(C, \mathbb{K}[C], \mathcal{F}_C)$, $(D, \mathbb{K}[D], \mathcal{F}_D)$ in the category of (pnc-)varieties.

The map

$$T_c\, pr_C \times T_d\, pr_D : \ T_{(c,d)}(C \times D) \ \to\ T_c C \times T_d D$$

is bijective. Explicitely, the inverse of $\delta_c \in T_c C$, $\delta_d \in T_d D$ is determined by

$$\begin{array}{rcll} \big((T_c\, pr_C \times T_d\, pr_D)^{-1}(\delta_c, \delta_d)\big)\,(f \otimes 1) & = & \delta_c(f) \quad , & f \in \mathbb{K}[C]\,, \\ \big((T_c\, pr_C \times T_d\, pr_D)^{-1}(\delta_c, \delta_d)\big)\,(1 \otimes g) & = & \delta_d(g) \quad , & g \in \mathbb{K}[D]\,. \end{array}$$

3. Principal open sets: Let $(A, \mathbb{K}[A], \mathcal{F}_A)$ be a (pnc-)variety, and let $g \in \mathbb{K}[A] \setminus \{0\}$. The localization $\mathbb{K}[A]_g$ topologized by the filter of ideals

$$\{ \; I^e \mid I \in \mathcal{F}_A \; \} \, , \quad \text{where} \quad I^e := \left\{ \left. \frac{f}{g^n} \; \right| \; f \in I \, , \; n \in \mathbb{N}_0 \right\} ,$$

gives in the obvious way a (pnc-)variety $(D(g), \mathbb{K}[D(g)], \mathcal{F}_{D(g)})$ on the principal open set $D(g) := \{ \; a \in A \mid g(a) \neq 0 \; \}$.

The inclusion map $j : D(g) \to A$ is a morphism, and for $a \in D(g)$ the tangent map $T_a j : T_a(D(g)) \to T_a A$ is bijective. Its inverse map is given explicitely by

$$(T_a j)^{-1}(\tilde{\delta})(\frac{f}{g^n}) \; = \; \frac{\tilde{\delta}(f) g(a)^n - f(a) \tilde{\delta}(g^n)}{g(a)^{2n}} \qquad , \qquad \tilde{\delta} \in T_a A \, , \; \frac{f}{g^n} \in \mathbb{K}[D(g)] \, .$$

Some propositions: The following two propositions on products and principal open sets of subvarieties will be important later:

Proposition 3.4. **(1)** *Let* $A_1, \ldots, A_k$ *be sub-pnc-varieties of a pnc-variety* B. *Let* $m : A_1 \times \ldots \times A_k \to B$ *be a surjective map, such that for every* $i \in \{1, 2, \ldots, k\}$ *and for every element* $a_i \in A_i$, *there exist elements* $a_j \in A_j$, $j \neq i$ *with* $m(a_1, \ldots, a_i, \ldots, a_k) = a_i$. *Suppose further that the comorphism* $m^* : \mathbb{K}[B] \to \mathbb{K}[A_1] \otimes \ldots \otimes \mathbb{K}[A_k]$ *exists and is surjective.*

Then m *and* m^* *are bijective, and* m^{-1} *is a morphism of pnc-varieties.*

(2) *Let* $A_1, \ldots, A_k$ *be varieties,* B *a pnc-variety, and* $\phi : B \to A_1 \times \ldots \times A_k$ *a bijective morphism with bijective comorphism* ϕ^*. *Then* B *is also a variety.*

PROOF. We first show part (1): Clearly m^* is injective, because m is surjective. To show the injectivity of m let $a_1, a_1' \in A_1, \ldots, a_k, a_k' \in A_k$, such that $m(a_1, \ldots, a_k) = m(a_1', \ldots, a_k')$. Let $i \in \{1, \ldots k\}$. For all $f \in \mathbb{K}[A_i]$ we have

$$f(a_i) \; = \; \Big((m^*)^{-1} (1 \otimes \cdots \otimes 1 \otimes \underbrace{f_i}_{i\text{-th place}} \otimes 1 \otimes \cdots \otimes 1) \Big) \Big(\underbrace{m(a_1, \ldots, a_k)}_{= m(a_1', \ldots, a_k')} \Big) \; = \; f(a_i') \, .$$

Because $\mathbb{K}[A_i]$ separates the points of A_i, we get $a_i = a_i'$.

It remains to show that m^{-1} is a morphism. Due to the definition of the filter of ideals for the product of the pnc-subvarieties, it is sufficient to show for $I \in \mathcal{F}_B$ and $i \in \{1, \ldots, k\}$ the inclusion

$$\Big((m^{-1})^* \Big)^{-1} (I) \; \supseteq \; 1 \otimes \ldots \otimes 1 \otimes \underbrace{I|_{A_i}}_{i\text{-th place}} \otimes 1 \otimes \ldots \otimes 1 \; .$$

Let $\tau_i : A_i \to B$ be the inclusion map, and define $\alpha_i := pr_i \circ m^{-1} : B \to A_i$. Because of $\alpha_i \circ \tau_i = id$ we find

$$\begin{aligned} \tau_i^*(I) \; &= \; ((\alpha_i \circ \tau_i)^*)^{-1} \left(\tau_i^*(I) \right) \; = \; (\alpha_i^*)^{-1} \Big((\tau_i^*)^{-1} \left(\tau_i^*(I) \right) \Big) \\ &= \; (\alpha_i^*)^{-1}(I) + \underbrace{(\alpha_i^*)^{-1} \left(\text{kernel} \, \tau_i^* \right)}_{= \{0\}} \; . \end{aligned}$$

Using this equation and the definition of α_i, we get

$$(m^{-1})^*\Big(1\otimes\ldots 1\otimes \underbrace{I|_{A_i}}_{i\text{-}th\ place} \otimes 1\ldots\otimes 1\Big)$$

$$= (m^{-1})^*\Big(pr_i^*\,(\tau_i^*(I))\Big) = \alpha_i^*\Big((\alpha_i^*)^{-1}(I)\Big) \subseteq I\ .$$

Now we prove part (2) of the proposition: We have to show that the map $B \to F\text{-Specm}\,\mathbb{K}\,[B]$ is surjective. Let $J \in F\text{-Specm}\,\mathbb{K}\,[B]$. Because ϕ is a morphism we have

$$(\phi^*)^{-1}(J) \in F\text{-Specm}\Big(\mathbb{K}\,[A_1]\otimes\ldots\otimes\mathbb{K}\,[A_k]\Big)\,.$$

$A_1\times\ldots\times A_k$ is a variety because $A_1, \ldots, A_k$ are varieties. Therefore there exist $a_1 \in A_1, \ldots, a_k \in A_k$ such that

$$(\phi^*)^{-1}(J) = I_{A_1\times\cdots\times A_k}\,((a_1, \ldots, a_k))\ .$$

Applying ϕ^* on both sides we get $J = I_B\left(\phi^{-1}(a_1, \ldots, a_k)\right)$.

□

Proposition 3.5. *Let $(B, \mathbb{K}\,[B], \mathcal{F}_B)$ be a variety and $g \in \mathbb{K}\,[B]$. Let $A \subseteq B$ be a sub-pnc-variety, such that $\overline{A} = B$.*

If $\left(D_A\,(g\,|_A), \mathbb{K}\,[\,D_A\,(g\,|_A)\,], \mathcal{F}_{D_A(g|_A)}\right)$ is a variety, then $D_A\,(g\,|_A) = D_B(g)$.

Proof. We only have to show $D_A\,(g\,|_A) \supseteq D_B(g)$. It is easy to see that the map

$$\phi:\ \mathbb{K}\,[B]_g \longrightarrow \mathbb{K}\,[A]_{g|_A}\quad ,\quad \frac{f}{g^n} \mapsto \frac{f\,|_A}{(g\,|_A)^n}$$

is a well defined, surjective homomorphism of algebras. It is injective because of $\overline{A} = B$. ϕ identifies the filter of ideals $\mathcal{F}_{D_B(g)}$ and $\mathcal{F}_{D_A(g|_A)}$, and therefore also $F\text{-Specm}\,\mathbb{K}\,[\,D_B(g)\,]$ and $F\text{-Specm}\,\mathbb{K}\,[\,D_A(g\,|_A)\,]$.

Now let $x \in D_B(g)$. Because $D_A(g\,|_A)$ is a variety, there exists an element $y \in D_A(g\,|_A)$ such that

$$\phi\left(I_{D_B(g)}(x)\right) = I_{D_A(g|_A)}(y)\,.$$

Applying ϕ^{-1} on both sides we find $I_{D_B(g)}(x) = I_{D_B(g)}(y)$, from which we conclude $x = y \in D_A(g\,|_A)$.

□

Let V be a vector space, and $(V, \mathbb{K}_X[V], \mathcal{F}_V)$ the variety given by a subspace $X \subseteq V^*$, which separates the points of V. Let $a \in A \subseteq V$.

- Then the Zariski closure of A is given by $\overline{A} = \mathcal{V}\,(I\,(A))$.
- Identify T_aA, $T_a\overline{A}$ with a subspace of T_aV using the tangential maps of the inclusions, and identify T_aV with V using the example of above. Then

$$T_a\overline{A} = T_aA = \left\{\ v \in V \ \middle|\ \delta_v\,(I\,(A)) = 0\,,\quad \begin{array}{c}\delta_v \text{ the derivation of } T_aV\\ \text{belonging to } v\,.\end{array}\right\}\,.$$

If the vanishing ideal $I\,(A)$ of A in $\mathbb{K}_X[V]$ is not known, but the coordinate ring $\mathbb{K}\,[A]$ is known very well, then nevertheless the Zariski closure and the tangent space can be determined:

Proposition 3.6. *We have:*

(a) $\overline{A} = \{ v \in V \mid \exists \beta_v \in \textit{Alg-Hom}(\mathbb{K}[A], \mathbb{K})\ \forall f \in X : \beta_v(f|_A) = f(v) \}$.
Furthermore for all $v, v' \in \overline{A}$: $\quad \beta_v = \beta_{v'} \iff v = v'$.

(b) $T_a\overline{A} = T_aA = \{ v \in V \mid \exists \epsilon_v \in Der_a\mathbb{K}[A]\ \forall f \in X : \epsilon_v(f|_A) = f(v) \}$.
Furthermore for all $v, v' \in T_a\overline{A}$: $\quad \epsilon_v = \epsilon_{v'} \iff v = v'$.

PROOF. We only show (a), the proof of (b) is analogous. $v \in \overline{A}$ is equivalent to $I(A) \subseteq I(v)$. This is equivalent to the fact, that the evaluation map $\tilde{\beta}_v : \mathbb{K}_X[V] \to \mathbb{K}$ in v factors to a homomorphism of algebras $\beta_v : \mathbb{K}[A] \to \mathbb{K}$, with $\beta_v(f|_A) = \tilde{\beta}_v(f) = f(v)$ for all $f \in X$.

If $\beta_v = \beta_{v'}$, then $v = v'$, because X separates the points of V. If $v = v'$, then $\beta_v = \beta_{v'}$, because $X|_A$ generates the algebra $\mathbb{K}[A]$.

□

3.2. Weak (pnc-)algebraic monoids

Let M be a monoid, and let $(M, \mathbb{K}[M], \mathcal{F}_M)$ be a (pnc-)variety, such that for all $m \in M$ the right and left translations r_m, l_m are morphisms of (pnc-)varieties. We get an injective linear map

$$\Psi_l : T_1M \to Der(\mathbb{K}[M])$$

by $(\Psi_l(\delta)f)(m) := \delta(l_m^* f)$, $f \in \mathbb{K}[M]$, $m \in M$. Its image is given by

$$\left\{ \epsilon \in \mathrm{Der}(\mathbb{K}[M]) \;\middle|\; \begin{array}{ll} \forall\, m \in M : & l_m^* \circ \epsilon = \epsilon \circ l_m^* \\ \exists\, J \in \mathcal{F}_M : & \epsilon(J) \subseteq I(1) \end{array} \right\} .$$

Similarly, we get an injective linear map

$$\Psi_r : T_1M \to Der(\mathbb{K}[M])$$

by $(\Psi_r(\delta)f)(m) := \delta(r_m^* f)$, $f \in \mathbb{K}[M]$, $m \in M$. Its image is given by

$$\left\{ \epsilon \in \mathrm{Der}(\mathbb{K}[M]) \;\middle|\; \begin{array}{ll} \forall\, m \in M : & r_m^* \circ \epsilon = \epsilon \circ r_m^* \\ \exists\, J \in \mathcal{F}_M : & \epsilon(J) \subseteq I(1) \end{array} \right\} .$$

The inverse maps on the images of Ψ_l and Ψ_r are obtained by concatenation with the evaluation morphism of $\mathbb{K}[M]$ in $1 \in M$.

We call $(M, \mathbb{K}[M], \mathcal{F}_M)$ a *weak (pnc-)algebraic monoid*, if the images of Ψ_l and Ψ_r are Lie subalgebras of $Der(\mathbb{K}[M])$, and if the Lie algebra structures on T_1M, obtained by pulling back, are opposite.

The tangent space T_1M equipped with the structure of the Lie algebra, which is obtained by pulling back by Ψ_l, is called the *Lie algebra* Lie(M) of M.

A *morphism* of weak (pnc-)algebraic monoids $(M, \mathbb{K}[M], \mathcal{F}_M)$, $(N, \mathbb{K}[N], \mathcal{F}_N)$ consists of a homomorphism of monoids $\phi : M \to N$, which is also a morphism of the (pnc-)varieties. The map

$$Lie(\phi) := T_1\phi : Lie(M) \to Lie(N)$$

is a homomorphism of Lie algebras.

Let $(M, \mathbb{K}[M], \mathcal{F}_M)$ be a weak (pnc-)algebraic monoid with an involution of monoids $* : M \to M$, which is also an isomorphism of (pnc-)varieties. This involution induces an involution $*$ of the algebra $\mathbb{K}[M]$ by $f^*(m) := f(m^*)$, $f \in \mathbb{K}[M]$, $m \in M$. It is easy to check that we also get an involution $*$ of the Lie algebra $Lie(M)$ by $\delta^* := \delta \circ *$, $\delta \in Lie(M)$.

Examples 3.7. In the following examples a dot $\cdot$ marks the place where the arguments of a function have to be put in.

(1) Let V be a vector space, and $(V, \mathbb{K}_X[V], \mathcal{F}_V)$ the variety given by a subspace $X \subseteq V^*$, which separates the points of V. For an endomorphism $\phi : V \to V$ denote by $\phi^* : V^* \to V^*$ its adjoint map. We get a subalgebra of $End(V)$ by

$$End_X(V) \quad := \quad \{ \phi \in End(V) \mid \phi^*(X) \subseteq X \} \ .$$

Equip $End_X(V)$ with the variety structure given by the point separating subspace

$$X_{End} \quad := \quad span\{ f_{\alpha v} := \alpha(\cdot v) \mid \alpha \in X \,,\, v \in V \}$$

of $(End_X(V))^*$. It is not difficult to check, that $End_X(V)$ together with the concatenation of linear maps is a weak algebraic monoid.

Identify $T_1 End_X(V)$ with $End_X(V)$ as in Example 3.3. Then the Lie algebra $Lie(End_X(V))$ identifies with the Lie algebra associated to the associative algebra $End_X(V)$.

(2) Let $(V\,,\,\langle\langle\ |\ \rangle\rangle)$ be a vector space together with a nondegenerate symmetric bilinear form, $char\,\mathbb{K} \neq 2$. Then

$$X \quad := \quad \{ f_v := \langle\langle v \mid \cdot \rangle\rangle \mid v \in V \} \ \subseteq \ V^*$$

is a subspace of V^*, which separates the points of V. The weak algebraic monoid $(\mathrm{End}_X(V), \mathbb{K}[End_X(V)], \mathcal{F}_{End_X(V)})$ of the last example coincides with

$$(\,\mathrm{Adj}(V)\,,\,\mathbb{K}[\,\mathrm{Adj}(V)\,]\,,\,\mathcal{F}_{Adj(V)}\,)$$

defined by

$$Adj(V) \quad := \quad \{ \phi \in End(V) \mid \ \text{The } \langle\langle\ |\ \rangle\rangle\text{-adjoint map } \phi^* \in End(V) \text{ exists.} \} \ ,$$
$$X_{Adj(V)} \quad := \quad \mathrm{span}\{ f_{vw} := \langle\langle v \mid \cdot\, w \rangle\rangle|_{Adj(V)} \mid v, w \in V \} \ \subseteq \ (Adj(V))^* \ .$$

The adjoint map $* : Adj \to Adj$ is an involutive isomorphism of varieties.

We omit the definition of an action of a weak (pnc-)algebraic monoid. But note that in Example 3.7, (1), for every $v \in V$ the map $\psi_v : End_X(V) \to V$ defined by $\psi_v(\phi) := \phi v$ is a morphism.

CHAPTER 4

A generalized Tannaka-Krein reconstruction

The category of representations $\mathcal{O}_{adm}$ of a Kac-Moody algebra $\mathbf{g}$ generalizes the category of finite dimensional representations of a semisimple Lie algebra, keeping the complete reducibility theorem. In this chapter we associate to $\mathcal{O}_{adm}$ and its category of restricted duals a monoid with coordinate ring, as well as the Lie algebra of this monoid, by an easy generalization of the Tannaka-Krein reconstruction. We show that this monoid can be identified with a weak algebraic monoid. Because of this result, we can use the algebraic geometric methods developed in the last chapter to determine this monoid and its Lie algebra. In the next two chapters we show, that the monoid can be identified with $\widehat{G}$, and its Lie algebra can be identified with the Kac-Moody algebra $\mathbf{g}$.

The generalized Tannaka-Krein reconstruction is not restricted to this particular example. It is a nice and systematic way to integrate infinite dimensional Lie algebras, which we will investigate in detail in [**M 4**]. But this example is special for the following reasons: There is a complete reducibility theorem for $\mathcal{O}_{adm}$. The restricted duals have nice properties. The Kac-Moody algebra is generated by elements, which act locally finite on all modules of $\mathcal{O}_{adm}$. Furthermore, a dense part of this monoid, the Kac-Moody group, is known very well. Also the restriction of the coordinate ring of the monoid onto the Kac-Moody group has been already investigated.

To each pair of $\mathbf{g}$-modules V, W of $\mathcal{O}_{adm}$ we assign a direct sum $V \oplus W$ and a tensor product $V \otimes W$, which are both modules of $\mathcal{O}_{adm}$. We also choose an one-dimensional module V_0 of $\mathcal{O}_{adm}$, on which $\mathbf{g}$ acts trivially.

The category of restricted duals of $\mathcal{O}_{adm}$ is a category of $\mathbf{g}^{op}$ modules. It is obtained as follows: Let (V, π_V) be a $\mathbf{g}$-module of $\mathcal{O}_{adm}$. Its restricted dual $V^{(*)}$ is defined as a linear space by

$$V^{(*)} := \bigoplus_{\lambda \in P(V)} V_\lambda^* \subseteq V^* .$$

It separates the points of V. For every $x \in \mathbf{g}$ the dual map $\pi_V(x)^* : V^* \to V^*$ of $\pi_V(x) : V \to V$ restricts to a map $\pi_V(x)^{(*)} : V^{(*)} \to V^{(*)}$, which we call the restricted dual map. In this way $V^{(*)}$ gets the structure of a $\mathbf{g}^{op}$-module.

For every $\mathbf{g}$-homomorphism $\phi : V \to W$ of $\mathcal{O}_{adm}$, the restricted dual map $\phi^{(*)} : W^{(*)} \to V^{(*)}$ exists, and is a homomorphism of $\mathbf{g}^{op}$-modules.

If V and W are $\mathbf{g}$-modules of $\mathcal{O}_{adm}$ we get in the obvious way linear embeddings

$$V^{(*)} \oplus W^{(*)} \to (V \oplus W)^* \quad , \quad V^{(*)} \otimes W^{(*)} \to (V \otimes W)^* .$$

Because the weight spaces are finite-dimensional, and because of the conditions on the set of weights of the modules of $\mathcal{O}_{adm}$, the images of these embeddings coincide with $(V \oplus W)^{(*)}$ and $(V \otimes W)^{(*)}$.

Let $\mathcal{V}ect$ be the category of $\mathbb{F}$-linear spaces and $fg : \mathcal{O}_{adm} \to \mathcal{V}ect$ be the functor forgetting the $\mathbf{g}$-module structure of the objects of $\mathcal{O}_{adm}$. A natural transformation of this functor consists of a family of linear maps

$$m \quad = \quad \big(m_V \in End(V) \big)_{V \; obj. \; of \; \mathcal{O}_{adm}} \; ,$$

such that the diagram

$$\begin{array}{ccc} V & \stackrel{m_V}{\to} & V \\ \phi \downarrow & & \downarrow \phi \\ W & \stackrel{m_W}{\to} & W \end{array}$$

commutes for all objects V, W, and all morphisms $\phi : V \to W$ of $\mathcal{O}_{adm}$. The natural transformations of fg form a set, because the isomorphism classes of $\mathcal{O}_{adm}$ form a set. Induced by the algebras of endomorphisms $End(V)$, V an object of $\mathcal{O}_{adm}$, the set of natural transformations of fg gets the structure of an associative $\mathbb{F}$-algebra with unit, therefore also the structure of a Lie algebra.

It is trivial to check the next proposition:

Proposition 4.1. *The set M of natural transformations*

$$m \quad = \quad \big(m_V \in End_{V^{(*)}}(V) \big)_{V \; obj. \; of \; \mathcal{O}_{adm}} \; ,$$

of the forgetful functor fg, which satisfy the properties

(1) $\quad m_{V \oplus W} = m_V \oplus m_W$ *for all objects V, W,*

(2) $\quad m_{V \otimes W} = m_V \otimes m_W$ *for all objects V, W,*

(3) $\quad m_{V_0} = id_{V_0}$,

is a submonoid of the monoid of natural transformations of fg.

Remarks. (1) M is the biggest monoid acting reasonably on the modules of $\mathcal{O}_{adm}$, compatible with the restricted duals.

(2) Recall that $m_V \in End_{V^{(*)}}(V)$ means the restricted dual map $m_V^{(*)} : V^{(*)} \to V^{(*)}$ exists. In this way $V^{(*)}$ is a module of the opposite monoid M^{op}.

(3) It is not necessary to demand (1). This follows immediately because m is a natural transformation.

More generally the following holds: Let V_j, $j \in J$, be objects of $\mathcal{O}_{adm}$, and $\bigoplus_{j \in J} V_j$ a sum of these representations. If this sum is also an object of $\mathcal{O}_{adm}$, then

$$m_{\bigoplus_{j \in J} V_j} \quad = \quad \bigoplus_{j \in J} m_{V_j} \; .$$

By Proposition 2.35 every element of $\widehat{G}$ gives a natural transformation of the forgetful functor fg. Therefore we get a morphism of monoids from $\widehat{G}$ into the monoid of natural transformations of fg. Now it is not difficult to check:

Proposition 4.2. *By this morphism the monoid $\widehat{G}$ can be identified with a submonoid of M.*

The next proposition describes the coordinate ring $\mathbb{F}[M]$ of the monoid M. For $\phi \in V^{(*)}$ and $v \in V$ define a function $f_{\phi v} : M \to \mathbb{F}$, the *matrix coefficient* of ϕ and v, by

$$f_{\phi v}(m) := \phi(m_V v) \quad , \quad m \in M .$$

Proposition 4.3. (1) *The set*

$$\mathbb{F}[M] := \left\{ f_{\phi v} \mid \phi \in V^{(*)} , v \in V , \ V \text{ an object of } \mathcal{O}_{adm} \right\}$$

is a subalgebra with unit of the algebra of functions on M. It separates the points of M. Left and right multiplications with elements of M induce comorphisms of $\mathbb{F}[M]$.

(2) *The monoid $M^{op} \times M$ acts on the algebra $\mathbb{F}[M]$ by*

$$((m_1, m_2)f)(m) := f(m_1 m m_2) \quad , \quad f \in \mathbb{F}[M] \quad , \quad m_1, m_2, m \in M .$$

We get an $M^{op} \times M$-equivariant linear isomorphism

$$\bigoplus_{\Lambda \in P^+} L(\Lambda)^{(*)} \otimes L(\Lambda) \to \mathbb{F}[M]$$

by assigning $\phi \otimes v$ the matrix coefficient $f_{\phi v}$, $\phi \in L(\Lambda)^{(*)}$, $v \in L(\Lambda)$, $\Lambda \in P^+$.

PROOF. (a) Let V, W be objects of $\mathcal{O}_{adm}$, and $\alpha : V \to W$ an isomorphism. Since the monoid M consists of natural transformations of the functor fg, we have

$$f_{\phi\, \alpha(v)} = f_{\alpha^{(*)}(\phi)\, v} \quad \text{for all} \quad \phi \in W^{(*)} \quad , \quad v \in V .$$

In particular the image of the map $V^{(*)} \otimes V \to \mathbb{F}[M]$, which assigns $\phi \otimes v$ the matrix coefficient $f_{\phi v}$, depends only on the isomorphism class of V.

(b) Next we show the properties of the set $\mathbb{F}[M]$, which are stated in (1): Let V, W be objects of $\mathcal{O}_{adm}$, and $v \in V$, $w \in W$, $\phi \in V^{(*)}$ and $\psi \in W^{(*)}$. Because of the properties (1) and (2) in the definition of M we have

$$f_{\phi v} + f_{\psi w} = f_{\phi \oplus \psi\, v \oplus w} \quad , \quad f_{\phi v} f_{\psi w} = f_{\phi \otimes \psi\, v \otimes w} .$$

Because of property (3) we have

$$f_{\phi v} = \phi(v)\, 1 \quad , \quad \phi \in V_0^{(*)} = V_0^* \quad , \quad v \in V_0 .$$

Let m, $\tilde{m} \in M$, and suppose $f(m) = f(\tilde{m})$ for all $f \in \mathbb{F}[M]$. Then for every object V of $\mathcal{O}_{adm}$ we have $\phi(m_V v) = \phi(\tilde{m}_V v)$ for all $\phi \in V^{(*)}$, $v \in V$. Because $V^{(*)}$ is point separating on V, we find $m_V = \tilde{m}_V$.

The left and right multiplications induce comorphisms, because for $v \in V$, $\phi \in V^{(*)}$, and V an object of $\mathcal{O}_{adm}$ we have

$$l_m^* f_{\phi v} = f_{m_V^{(*)}(\phi)\, v} \quad , \quad r_m^* f_{\phi v} = f_{\phi\, m_V(v)} .$$

(c) We get a linear map

$$\Phi_M : \bigoplus_{\Lambda \in P^+} L(\Lambda)^{(*)} \otimes L(\Lambda) \to \mathbb{F}[M]$$

by assigning $\phi \otimes v$ the matrix coefficient $f_{\phi v}$, $\phi \in L(\Lambda)^{(*)}$, $v \in L(\Lambda)$, $\Lambda \in P^+$.

This map is surjective: Let V be an object of $\mathcal{O}_{adm}$. Due to the complete reducibility theorem, [**K**], Theorem 10.7, there exists an isomorphism $\alpha : V \to \bigoplus_{j \in J} L(\Lambda_j)$, where $\Lambda_j \in P^+$. Let $v \in V$ and $\phi \in V^{(*)}$. Since $\alpha^{(*)} : \bigoplus_{j \in J} L(\Lambda_j)^{(*)} \to$

$V^{(*)}$ is bijective, there exist an element $\psi \in \bigoplus_{j\in J} L(\Lambda_j)^{(*)}$ such that $\phi = \alpha^{(*)}(\psi)$. Due to part (a) we find

$$f_{\phi v} \;=\; f_{\alpha^{(*)}(\psi)\, v} \;=\; f_{\psi\, \alpha(v)} \;.$$

This matrix coefficient is an element of the image of Φ_M.

Denote by $\Phi_G : \bigoplus_{\Lambda\in P^+} L(\Lambda)^{(*)} \otimes L(\Lambda) \to \mathbb{F}[G]$ the linear isomorphism of the Peter and Weyl theorem for the algebra of strongly regular functions $\mathbb{F}[G]$. Identify the Kac-Moody group G with its corresponding subgroup of M. Obviously we get a linear map $res : \mathbb{F}[M] \to \mathbb{F}[G]$ by restricting the functions of $\mathbb{F}[M]$ onto G. Because of $\Phi_G = res \circ \Phi_M$, the map Φ_M is injective.

The monoid $M^{op} \times M$ acts by homomorphism of algebras, because of

$$(m_1, m_2)f \;=\; (l^*_{m_1} \circ r^*_{m_2})(f) \quad , \quad m_1, m_2 \in M \; , \quad f \in \mathbb{F}[M] \; .$$

The $M^{op} \times M$-equivariance is obvious.

□

Corollary 4.4. **(1)** *Identify the monoid $\widehat{G}$ with the corresponding submonoid of M. Then:*

- *The Kac-Moody group G is Zariski closed in M.*
- *The algebra of strongly regular functions $\mathbb{F}[G]$ is isomorphic to the coordinate ring $\mathbb{F}[M]$ by the restriction map.*

(2) *The coordinate ring $\mathbb{F}[M]$ is an integrally closed domain. In particular M is irreducible.*

Remark: In this way, the algebra of strongly regular functions is really the coordinate ring of the monoid M.

PROOF. With the notations of the proof of the last proposition, the restriction map $res = \Phi_G \circ \Phi_M^{-1}$ is bijective. From this follows part (1) of the corollary.

Since $\mathbb{F}[M]$ is isomorphic the algebra of strongly regular functions, part (2) follows from the properties of the algebra of strongly regular functions.

□

The next Proposition gives the Lie algebra $Lie(M)$ of $(M,\ \mathbb{F}[M])$, and the adjoint action of the unit group $M^\times$ on $Lie(M)$.

Proposition 4.5. *Let $Lie(M)$ be the set of natural transformations*

$$x \;=\; \left(x_V \in End_{V^{(*)}}(V)\right)_{V \text{ obj. of } \mathcal{O}_{adm}}$$

of the forgetful functor fg, which satisfy the following properties:

(1) $x_{V\oplus W} = x_V \oplus x_W$ *for all objects V, W.*

(2) $x_{V\otimes W} = x_V \otimes id_W + id_V \otimes x_W$ *for all objects V, W.*

(3) $x_{V_0} = 0_{V_0}$.

(4) *There exists a derivation $\delta_x : \mathbb{F}[M] \to \mathbb{F}$ in $1 \in M$, such that*

$$\delta_x(f_{\phi v}) \;=\; \phi(x_V v) \;\; \text{for all } \phi \in V^{(*)},\; v \in V,\; V \text{ an object of } \mathcal{O}_{adm} \; .$$

Then Lie(M) is a Lie subalgebra of the Lie algebra of natural transformations of the forgetful functor fg. The unit group $M^\times$ of M acts on $Lie(M)$ by conjugation.

Remarks. (1) The Lie algebra $Lie(M)$ is the biggest Lie algebra acting reasonably on the modules of $\mathcal{O}_{adm}$, compatible with the restricted duals, and with the coordinate ring $\mathbb{F}[M]$.

(2) Clearly we want to have the compatibility condition, that the Lie algebra of natural transformations, corresponding to the Kac-Moody algebra $\mathbf{g}$, is a subalgebra of $Lie(M)$. We will show in the last chapter that we have even equality.

PROOF. Let $x, y \in Lie(M)$ and $m \in M^\times$. It is trivial to check, that (1), (2), and (3) are also valid for $[x,y]$ and mxm^{-1}. Note that statement (4) is equivalent to the statement:

(4') There exists a derivation $\tilde{\delta}_x \in Der(\mathbb{F}[M])$ such that

$$\tilde{\delta}_x(f_{\phi v}) = f_{\phi\, x_V v} \text{ for all } \phi \in V^{(*)},\ v \in V,\ V \text{ an object of } \mathcal{O}_{adm}\ .$$

The derivation $\tilde{\delta}_x$ is the left invariant derivation of $\mathbb{F}[M]$ corresponding to δ_x.

If $x, y \in Lie(M)$, then $[\tilde{\delta}_x, \tilde{\delta}_y] \in Der(\mathbb{F}[M])$ has the properties of (4') for $[x,y]$. For $m \in M^\times$ the comorphisms l_m^*, $r_{m^{-1}}^*$ of the left and right multiplications l_m, $r_{m^{-1}}$ by m, m^{-1} are isomorphism of algebras. If $x \in Lie(M)$, then $\delta_x \circ l_m^* \circ r_{m^{-1}}^*$ has the properties of (4) for mxm^{-1}.

□

Recall the weak algebraic monoid of Example 3.7, (1). Now regard

$$End_X\left(\bigoplus_{\Lambda\in P^+} L(\Lambda)\right) \quad \text{with} \quad X := \bigoplus_{\Lambda\in P^+} L(\Lambda)^{(*)} \subseteq \left(\bigoplus_{\Lambda\in P^+} L(\Lambda)\right)^* .$$

It is easy to check, that

$$gr\text{-}End_X\left(\bigoplus_{\Lambda\in P^+} L(\Lambda)\right)$$
$$:= \left\{ \phi \in End_X\left(\bigoplus_{\Lambda\in P^+} L(\Lambda)\right) \;\middle|\; \phi\,(L(\Lambda)) \subseteq L(\Lambda),\ \Lambda \in P^+ \right\}$$

is a closed submonoid. We will work with this weak algebraic monoid for the following reason: Equip $L(\Lambda)$ with the variety structure given by

$$L(\Lambda)^{(*)} \subseteq L(\Lambda)^*\ , \quad \Lambda \in P^+\ .$$

$End_X\left(\bigoplus_{\Lambda\in P^+} L(\Lambda)\right)$ does not act on $L(\Lambda)$, but $gr\text{-}End_X\left(\bigoplus_{\Lambda\in P^+} L(\Lambda)\right)$ acts, and for every $v \in L(\Lambda)$ the map

$$\begin{array}{ccc} gr\text{-}End_X\left(\bigoplus_{\Lambda\in P^+} L(\Lambda)\right) & \to & L(\Lambda) \\ \phi & \mapsto & \phi v \end{array}$$

is a morphism of varieties, $\Lambda \in P^+$.

By definition $\widehat{G} \subseteq End\left(\bigoplus_{\Lambda\in P^+} L(\Lambda)\right)$. Actually we have

$$\widehat{G} \subseteq gr\text{-}End_X\left(\bigoplus_{\Lambda\in P^+} L(\Lambda)\right) .$$

In particular the closure $\overline{G} \subseteq gr\text{-}End_X\left(\bigoplus_{\Lambda\in P^+} L(\Lambda)\right)$ gets the structure of a weak algebraic monoid.

Theorem 4.6. *We get an embedding of monoids*

$$\Upsilon : M \quad \to \quad gr\text{-}End_X\left(\bigoplus_{\Lambda\in P^+} L(\Lambda)\right) \quad ,$$

with image $\overline{G}$, by

$$\Upsilon(m)v \;:=\; m_{L(\Lambda)}v \quad , \qquad m\in M\,,\; v\in L(\Lambda)\,,\; \Lambda\in P^+\,.$$

If we identify M with $\overline{G}$ by Υ, then the coordinate ring $\mathbb{F}[M]$ identifies with the coordinate ring $\mathbb{F}[\overline{G}]$, and the Lie algebra $Lie(M)$ identifies with $Lie(\overline{G})$ by the map $x \mapsto \delta_x$.

Remark. By Proposition 4.2 the monoid $\widehat{G} \subseteq gr\text{-}End_X\left(\bigoplus_{\Lambda\in P^+} L(\Lambda)\right)$ embeds into M. If we restrict to the image of this embedding, then the inverse map is given by Υ.

PROOF. First we show $\Upsilon(M) \subseteq \overline{G}$. Every element $m \in M$ determines an evaluation homomorphism of $\mathbb{F}[M]$. Due to Corollary 4.4 the coordinate ring $\mathbb{F}[M]$ is isomorphic to the algebra of strongly regular functions $\mathbb{F}[G]$ by the restriction map. Therefore we also get an algebra homomorphism $\beta_m : \mathbb{F}[G] \to \mathbb{F}$, which has the property

$$\beta_m(f_{\phi v}|_G) \;=\; \phi(m_{L(\Lambda)}v) \quad \text{for all} \quad \phi\in L(\Lambda)^{(*)}\,,\; v\in L(\Lambda)\,,\; \Lambda\in P^+\,.$$

Due to Proposition 3.6 we find $\Upsilon(m) \in \overline{G}$.

Denote by $\tilde{\Upsilon} : M \to \overline{G}$ the map Υ, whose image has been restricted to $\overline{G}$. We first show that this map is surjective by constructing a map $\Omega : \overline{G} \to M$ with the property $\tilde{\Upsilon}\circ\Omega = id_{\overline{G}}$:

Let $\widehat{g} \in \overline{G}$. For every object V of $\mathcal{O}_{adm}$ choose an isomorphism

$$\gamma : V \;\to\; \bigoplus_{j\in J} L(\Lambda_j)\,,$$

and define a linear map $\Omega_V(\widehat{g}) := \gamma^{-1}\circ\widehat{g}\circ\gamma$. Because the adjoint maps of γ and $\widehat{g}$ exist, we have $\Omega_V \in End_{V^{(*)}}(V)$. The next formula shows that the map $\Omega_V(\widehat{g})$ is also independent of the chosen isomorphism.

The algebra $\mathbb{F}[G]$ is isomorphic to $\mathbb{F}[\overline{G}]$ by the restriction map. Denote by $\alpha_{\widehat{g}} : \mathbb{F}[G] \to \mathbb{F}$ the homomorphism, which corresponds to the evaluation homomorphism of $\mathbb{F}[\overline{G}]$ in $\widehat{g}$. It is easy to check that we have

$$\phi(\Omega_V(\widehat{g})v) \;=\; \alpha_{\widehat{g}}(f_{\phi v}|_G) \quad \text{for all} \quad \phi\in V^{(*)} \quad , \quad v\in V\;.$$

Using this formula, it is also not difficult to see, that the maps $\Omega_V(\widehat{g})$, V an object of $\mathcal{O}_{adm}$, define an element $\Omega(\widehat{g})$ of M. Obviously $\tilde{\Upsilon}(\Omega(\widehat{g})) = \widehat{g}$.

It is easy to check, that the comorphism $\tilde{\Upsilon}^* : \mathbb{F}[\overline{G}] \to \mathbb{F}[M]$ exists, and is given by the concatenation of the linear isomorphisms

$$\mathbb{F}[\overline{G}] \overset{res}{\to} \mathbb{F}[G] \overset{\Phi_G^{-1}}{\to} \bigoplus_{\Lambda\in P^+} L(\Lambda)^{(*)}\otimes L(\Lambda) \overset{\Phi_M}{\to} \mathbb{F}[M] \;,$$

in particular $\tilde{\Upsilon}^*$ is an isomorphism.

The map $\tilde{\Upsilon}$ is also injective. Let $m, m' \in M$, such that $\tilde{\Upsilon}(m) = \tilde{\Upsilon}(m')$. Then for all $\phi \in L(\Lambda)^{(*)}$, $v \in L(\Lambda)$, and $\Lambda \in P^+$ we have

$$\tilde{\Upsilon}^*(f_{\phi v})(m) \;=\; \phi(\tilde{\Upsilon}(m)v) \;=\; \phi(\tilde{\Upsilon}(m')v) \;=\; \tilde{\Upsilon}^*(f_{\phi v})(m') \,.$$

Due to Proposition 4.3 the functions of $\mathbb{F}[M]$ separate the points of M. Because $\tilde{\Upsilon}^*$ is surjective, we find $m = m'$.

Identify $Lie(\overline{G})$ with a subalgebra of the Lie algebra $gr\text{-}End_X\left(\bigoplus_{\Lambda\in P^+} L(\Lambda)\right)$. It remains to show that we get an embedding of Lie algebras

$$\Upsilon' : \, Lie(M) \;\to\; gr\text{-}End_X\left(\bigoplus_{\Lambda\in P^+} L(\Lambda)\right) \,,$$

with image $Lie(\overline{G})$, by

$$\Upsilon'(x)v \;:=\; x_{L(\Lambda)}v \qquad , \qquad x \in Lie(M)\,,\; v \in L(\Lambda)\,,\; \Lambda \in P^+ \,.$$

This can proved in a similar way as the corresponding result for Υ.

□

Recall that we fixed a nondegenerate contravariant symmetric bilinear form $\langle\langle \mid \rangle\rangle$ on every module $L(\Lambda)$, $\Lambda \in P^+$. We extended these forms to a form on $\bigoplus_{\Lambda\in P^+} L(\Lambda)$, also denoted by $\langle\langle \mid \rangle\rangle$, by requiring $L(\Lambda)$ and $L(\Lambda')$ to be orthogonal for $\Lambda \neq \Lambda'$. It will be easier to work with some of the last definitions and results, if we rewrite them, using these forms:

As above, in the following definitions a dot $\cdot$ marks the place where the arguments of a function have to be put in.

The variety structure on $L(\Lambda)$ of above is given by

$$L(\Lambda)^{(*)} \;=\; \{\; f_v := \langle\langle v \mid \cdot \,\rangle\rangle \;\mid\; v \in L(\Lambda) \;\} \;\subseteq\; (L(\Lambda))^* \,.$$

The weak algebraic monoid $gr\text{-}End_X\left(\bigoplus_{\Lambda\in P^+} L(\Lambda)\right)$ can be identified with

$$gr\text{-}Adj\left(\bigoplus_{\Lambda\in P^+} L(\Lambda)\right)$$
$$:= \left\{\; \phi \in End\left(\bigoplus_{\Lambda\in P^+} L(\Lambda)\right) \;\middle|\; \begin{array}{l}\text{The adjoint } \phi^* \text{ exists and} \\ \phi\,(L(\Lambda)) \subseteq L(\Lambda),\; \Lambda \in P^+ \,.\end{array} \right\} \,,$$

whose structure of a variety is given by

$$\mathrm{span}\left\{\; f_{vw} := \langle\langle v \mid \cdot\; w\rangle\rangle\,|_{gr\text{-}Adj} \;\mid\; v, w \in L(\Lambda)\,,\; \Lambda \in P^+ \;\right\} \;.$$

The adjoint map $*$ of $gr\text{-}Adj\left(\bigoplus_{\Lambda\in P^+} L(\Lambda)\right)$ is an involutive isomorphism. It restricts to the Chevalley involution on $\widehat{G} \subseteq gr\text{-}Adj\left(\bigoplus_{\Lambda\in P^+} L(\Lambda)\right)$. Since $\widehat{G}$ is Zariski dense in $\overline{G}$, also the closure $\overline{G}$ is invariant under $*$. We call the adjoint map $*$ on $\overline{G}$ also *Chevalley involution.*

We use the Chevalley involution to get rid of the $\overline{G}^{op} \times \overline{G}$-action on the coordinate ring of $\overline{G}$: Define an action π of $\overline{G} \times \overline{G}$ on $\mathbb{F}[\overline{G}]$, and an involution $*$ of $\mathbb{F}[\overline{G}]$, which we also call *Chevalley involution*, by

$$\begin{array}{rcl} \pi(x,y)\,f &:=& f(x^* \cdot y) \\ f^*(x) &:=& f(x^*) \end{array} \;, \quad \text{where} \quad x, y \in G\,, \quad f \in \mathbb{F}[\overline{G}] \;.$$

There is the following *Peter and Weyl theorem* : We get a $\overline{G}\times\overline{G}$-equivariant linear bijective map

$$\Psi : \bigoplus_{\Lambda\in P^+} L(\Lambda)\otimes L(\Lambda) \ \to \ \mathbb{F}\left[\overline{\mathrm{G}}\right]$$

by assigning

$$v\otimes w \quad \text{the function} \quad f_{vw}\,|_{\overline{G}} := \langle\langle v \mid \cdot\, w\rangle\rangle\,|_{\overline{G}}\,,\ v,w\in L(\Lambda)\,,\ \Lambda\in P^+\,.$$

It identifies the sum of the switch-maps with the Chevalley involution.

Remark. Due to the last theorem, our main aims in the next two chapters are to determine the closure $\overline{G}\subseteq gr\text{-}Adj\left(\bigoplus_{\Lambda\in P^+} L(\Lambda)\right)$, as well as the Lie algebra $Lie(\overline{G})$ of the weak algebraic monoid $\overline{G}$.

CHAPTER 5

The proof of $\overline{G} = \widehat{G}$ and some other theorems

In this chapter we prove that the closure $\overline{G} \subseteq gr\text{-}Adj\left(\bigoplus_{\Lambda \in P^+} L(\Lambda) \right)$ coincides with $\widehat{G}$. As explained in the last chapter this shows that $\widehat{G}$ is the monoid associated to the category $\mathcal{O}_{adm}$ and its category of restricted duals.

We also determine some other closures. In particular we show $\overline{T} = \widehat{T}$. Furthermore, $\overline{T}$ is a generalized affine toric variety, and the action of the Weyl group on T extends to an action on $\overline{T}$. We show that $U^{\pm}$ is closed. We show $\overline{N} = \widehat{N}$, which makes the analogy of the Weyl monoid $\widehat{\mathcal{W}} \cong \widehat{N}/T = \overline{N}/T$ with a Renner monoid even more close. We also show that the dense unit group G is open in $\overline{G}$.

Combined with the structural results obtained in Chapter 2, these results indicate that the monoid $\overline{G} = \widehat{G}$, together with its coordinate ring, is really an analogue of a reductive algebraic monoid.

Although $\overline{G}$ does not act on the flag varieties realized in the projective spaces $\mathbb{P}(L(\Lambda))$, $\Lambda \in P^+$, we show that it acts on the corresponding affine cones.

We show that the Kac-Peterson-Slodowy part of the $\mathbb{F}$-valued points of the algebra of strongly regular functions coincides with the part, which is given by $\widehat{G}$.

The hearts of the proofs of $\overline{G} = \widehat{G}$ and $\overline{N} = \widehat{N}$ are inductions over the cardinality of $J \subseteq I$, showing $\overline{G_J} = \widehat{G}_J$ and $\overline{N_J} = \widehat{N}_J$. The results in the next sections, describing something indexed by J, or indexed by '*rest*', are needed to prepare these proofs.

5.1. The coordinate rings and closures of T, T_J $(J \subseteq I)$, and T_{rest}

First some notations and remarks. Let $J \subseteq I$. There are the following sublattices of H and P:

$$\begin{array}{llllll} H_J & := & \mathbb{Z}\text{-span}\{\, h_j \mid j \in J \,\}\ , & H_{rest} & := & \{\, h \in H \mid \Lambda_i(h) = 0 \ \text{ for } \ i \in I \,\}\ , \\ P_J & := & \mathbb{Z}\text{-span}\{\, \Lambda_j \mid j \in J \,\}\ , & P_{rest} & := & \{\, \lambda \in P \mid \lambda(h_i) = 0 \ \text{ for } \ i \in I \,\}\ . \end{array}$$

Define the following projections:

$$\begin{array}{rccc} p_J : & P & \to & P_J \\ & \lambda & \mapsto & \sum_{j \in J} \lambda(h_j)\Lambda_j \end{array} \ , \qquad \begin{array}{rccc} p_{rest} : & P & \to & P_{rest} \\ & \lambda & \mapsto & \lambda - p_I(\lambda) \end{array} \ .$$

Recall that the torus T of the Kac-Moody group can be described by the following two isomorphisms of groups:

$$\begin{array}{ccc} H \otimes_{\mathbb{Z}} \mathbb{F}^{\times} & \to & T \\ \sum_\tau h_\tau \otimes s_\tau & \mapsto & \prod_\tau t_{h_\tau}(s_\tau) \end{array} \ , \qquad \begin{array}{ccc} \mathrm{Hom}(P, \mathbb{F}^{\times}) & \to & T \\ \alpha & \mapsto & t(\alpha) \end{array} \ ,$$

where $t_h(s)v_\lambda := s^{\lambda(h)}v_\lambda$ and $t(\alpha)v_\lambda := \alpha(\lambda)v_\lambda$, $v_\lambda \in L(\Lambda)_\lambda$, $\Lambda \in P^+$.

Recall that the group algebra $\mathbb{F}[P]$ of the lattice P can be identified with the classical coordinate ring of the torus T, identifying $\sum c_\lambda e_\lambda \in \mathbb{F}[P]$ with the function

$$\left(\sum_\lambda c_\lambda\, e_\lambda\right)(\,t(\alpha)\,) \;:=\; \sum_\lambda c_\lambda\, \alpha(\lambda) \qquad , \qquad \alpha \in \mathrm{Hom}(P, \mathbb{F}^\times)\,.$$

We have similar descriptions for the torus T_J, replacing T, H, P by T_J, H_J, P_J, where now $t(\alpha)$ acts as

$$t(\alpha)\, v_\lambda \;:=\; \alpha(p_J(\lambda))\, v_\lambda \qquad , \qquad v_\lambda \in L(\Lambda)_\lambda\,,\;\; \Lambda \in P^+\,.$$

We have similar descriptions for the torus T_{rest}.

To determine the closures of T_J, T, and T_{rest} we need the coordinate rings $\mathbb{F}[T_J]$, $\mathbb{F}[T]$, and $\mathbb{F}[T_{rest}]$. Except the coordinate ring of T_{rest}, these are in general only subalgebras of the classical coordinate rings:

Proposition 5.1. *Let $J \subseteq I$. We have:*

(1) $\mathbb{F}[T_J] \;=\; \mathbb{F}[p_J(X\cap P)] \;=\; \mathbb{F}[p_{J^\infty}(X\cap P) + P_{J^0}]\,.$

(2) $\mathbb{F}[T] \;=\; \mathbb{F}[X\cap P]\,.$

(3) $\mathbb{F}[T_{rest}] \;=\; \mathbb{F}[P_{rest}]\,.$

PROOF. We only show (1) in the nontrivial case $J \neq \emptyset$. The proofs of (2) and (3) are similar.

(a) Let $v_\lambda \in L(\Lambda)_\lambda$, $w_\mu \in L(\Lambda)_\mu$, $\lambda, \mu \in P(\Lambda)$, and $\Lambda \in P^+$. By checking on the elements of T_J, we find

$$f_{v_\lambda w_\mu}|_{T_J} \;=\; \langle\langle v_\lambda \mid w_\mu \rangle\rangle\, e_{p_J(\mu)} \;=\; f_{v_\lambda w_\mu}(1)\, e_{p_J(\mu)}\,. \tag{5.1}$$

Due to the Peter and Weyl theorem for $\mathbb{F}[G]$, and $\bigcup_{\Lambda\in P^+} P(\Lambda) = X \cap P$, we get $\mathbb{F}[T_J] \subseteq \mathbb{F}[p_J(X\cap P)]$. Due to the nondegeneracy of the forms $\langle\langle\ \mid\ \rangle\rangle$ on the weight spaces, we have even equality.

(b) Let $J^0 \neq \emptyset$. We have

$$p_J(X\cap P) \;=\; (p_{J^\infty} + p_{J^0})(X\cap P) \;\subseteq\; p_{J^\infty}(X\cap P) + P_{J^0}\,.$$

To show the reverse inclusion, note that the comorphism of the multiplication map $G_{J^\infty} \times G_{J^0} \to G_J$ gives an isomorphism of algebras $\mathbb{F}[G_J] \to \mathbb{F}[G_{J^\infty}] \otimes \mathbb{F}[G_{J^0}]$. It is easy to conclude, that the comorphism of the multiplication map $m : T_{J^\infty} \times T_{J^0} \to T_J$ exists, and gives an isomorphism of algebras

$$m^* : \mathbb{F}[T_J] \;\;\to\;\; \mathbb{F}[T_{J^\infty}] \otimes \mathbb{F}[T_{J^0}]\,.$$

Due to (a) we have $\mathbb{F}[T_J] = \mathbb{F}[p_J(X\cap P)]$ and $\mathbb{F}[T_{J^\infty}] = \mathbb{F}[p_{J^\infty}(X\cap P)]$. Because G_{J^0} is a Kac-Moody group of finite type with maximal torus T_{J^0}, we have $\mathbb{F}[T_{J^0}] = \mathbb{F}[P_{J^0}]$. By checking on the elements of T_J, we find

$$(m^{-1})^*\,(e_{\lambda_\infty} \otimes e_{\lambda_0}) \;=\; e_{\lambda_\infty + \lambda_0} \quad , \quad \lambda_\infty \in p_{J^\infty}(X\cap P)\,,\; \lambda_0 \in P_{J^0}\,.$$

Therefore $p_{J^\infty}(X\cap P) + P_{J^0} \subseteq p_J(X\cap P)$.

□

Theorem 5.2. *Let $J \subseteq I$. We have:*

(1) $\overline{T_J} \;=\; \widehat{T}_J\,.$

(2) $\overline{T} \;=\; \widehat{T}\,.$

(3) T_{rest} *is closed.*

PROOF. We only show (1) in the nontrivial case $J \neq \emptyset$. The proofs of (2) and (3) are similar to part (c) of the following proof.

(a) To prepare the proof, we show that the submonoid $p_J(X \cap P)$ of P_J is saturated, and its faces are given by

$$Fa\,(p_J(X \cap P)) \;=\; \{\; p_J(R \cap P) \;|\; R \in \mathcal{R}(X)_J \;\} \;.$$

• We first treat the case $J = J^\infty$. Due to Theorem 2.2 (c), the kernel of p_{J^∞} is the hull subgroup of $R(J^\infty) \cap P$, i.e.,

$$\text{kernel}\; p_{J^\infty} \;=\; \big(R(J^\infty) - R(J^\infty)\big) \cap P \,. \tag{5.2}$$

Recall that $(X - R(J^\infty)) \cap P$ is a saturated submonoid of P, whose faces are given by

$$\begin{aligned} Fa(\,(X - R(J^\infty)) \cap P\,) \;&=\; \{\; (R - R(J^\infty)) \cap P \;|\; R \in \mathcal{R}(X)\,,\; R \supseteq R(J^\infty)\;\} \\ &=\; \{\; (R - R(J^\infty)) \cap P \;|\; R \in \mathcal{R}(X)_J \;\} \;. \end{aligned}$$

The surjective linear map $p_{J^\infty} : P \to P_{J^\infty}$ maps the saturated submonoid $(X - R(J^\infty)) \cap P$ surjectively onto $p_{J^\infty}(X \cap P)$. Its kernel (5.2) is smallest the face of $(X - R(J^\infty)) \cap P$.

Using these facts, it is easy to check, that $p_{J^\infty}(X \cap P)$ is saturated. It is also not difficult to check, that the map

$$\begin{array}{ccc} Fa(\,(X - R(J^\infty)) \cap P\,) & \to & Fa(\,p_{J^\infty}(X \cap P)\,) \\ F & \mapsto & p_{J^\infty}(F) \end{array}$$

is bijective. Using the description of the faces of $(X - R(J^\infty)) \cap P$, and once more (5.2), we get the required result.

• For an arbitrary $J \subseteq I$, we have due to the last proposition:

$$p_J(X \cap P) \;=\; p_{J^\infty}(X \cap P) + P_{J^0} \;\subseteq\; P_{J^\infty} \oplus P_{J^0} \;. \tag{5.3}$$

Using the preceeding results, it is easy to see, that $p_J(X \cap P)$ is saturated in P_J, and its faces are given by

$$Fa\,(p_J(X \cap P)) \;=\; \{p_{J^\infty}(R \cap P) + P_{J^0} \;|\; R \in \mathcal{R}(X)_{J^\infty} = \mathcal{R}(X)_J\,\} \,.$$

Because of $p_J = p_{J^\infty} + p_{J^0}$ and $p_{J^\infty} = p_J - p_{J^0}$, we find

$$p_J(R \cap P) \;\subseteq\; p_{J^\infty}(R \cap P) + P_{J^0} \;\subseteq\; p_J(R \cap P) + P_{J^0} \;\;.$$

To show that the first inclusion is an equality, it is sufficient to show that $P_{J^0} \subseteq p_J(R \cap P)$. Let $\mu \in P_{J^0}$. Due to (5.3) there exists an element $\lambda \in X \cap P$, such that $\mu = p_J(\lambda)$. From this follows

$$0 \;=\; \mu(h_i) = p_J(\lambda)(h_i) \;=\; \lambda(h_i) \qquad \text{for all} \qquad i \in J^\infty \;.$$

Therefore

$$\lambda \;\in\; X \cap (R(J^\infty) + R(J^\infty)) \cap P \;=\; R(J^\infty) \cap P\,,$$

and $\mu \in p_J(R(J^\infty) \cap P) \subseteq p_J(R \cap P)$.

(b) Fix $\mu \in X \cap P$ and $R \in \mathcal{R}(X)_J$. Next we show

$$\mu \;\in\; R \cap P \quad\Longleftrightarrow\quad p_J(\mu) \;\in\; p_J(R \cap P) \;\;.$$

The direction '$\Rightarrow$' is trivial. Now let $p_J(\mu) \in p_J(R \cap P)$. By using (5.2) we find

$$\begin{aligned} \mu \;\in\; ((R \cap P) + \text{kernel}\; p_J) \cap X \;&\subseteq\; ((R \cap P) + \text{kernel}\; p_{J^\infty}) \cap X \\ &=\; (R - R(J^\infty)) \cap X \cap P \,. \end{aligned}$$

Suppose there exist elements $r_1 \in R$, $r_2 \in R(J^\infty)$ with $r_1 - r_2 \in X \setminus R$. Because R is a face of X, we get $r_1 \in X \setminus R + r_2 \subseteq X \setminus R$, which is a contradiction. Therefore we conclude $(R - R(J^\infty)) \cap X = R$.

(c) Due to Proposition 3.6 the Zariski closure of T_J is given by

$$\overline{T_J} = \Bigg\{ \phi \in gr\text{-}Adj\left(\bigoplus_{\Lambda \in P^+} L(\Lambda)\right) \;\Bigg|\; \exists \text{ hom. of alg. } \beta_\phi : \mathbb{F}[T_J] \to \mathbb{F}$$

$$\forall\, v, w \in L(\Lambda),\, \Lambda \in P^+ : \quad \beta_\phi(f_{vw}|_{T_J}) = \langle\langle v \mid \phi w \rangle\rangle \Bigg\} ,$$

and $\phi \in \overline{T_J}$ is uniquely determined by β_ϕ. Due to the last proposition we have $\mathbb{F}[T_J] = \mathbb{F}[p_J(X \cap P)]$.

For $t \in T_J$ let $\alpha_t \in \mathrm{Hom}(P_J, \mathbb{F}^\times)$ be the corresponding homomorphism of monoids. Note that $p_J(X \cap P)$ spans P_J, because $X \cap P$ spans P. Due to (a) and Remark 1.5 we get for every $t \in T_J$ and $R \in \mathcal{R}(X)_J$ a homomorphisms of algebras $\beta_{t,R} : \mathbb{F}[p_J(X \cap P)] \to \mathbb{F}$ by

$$\beta_{t,R}(e_\lambda) := \begin{cases} \alpha_t(\lambda) & \text{if} \quad \lambda \in p_J(R \cap P) \\ 0 & \text{if} \quad \lambda \in p_J(X \cap P) \setminus p_J(R \cap P) \end{cases} .$$

Using equation (5.1) and (b) we find for all $v_\lambda \in L(\Lambda)_\lambda$, $w_\mu \in L(\Lambda)_\mu$, $\lambda, \mu \in P(\Lambda)$ and $\Lambda \in P^+$:

$$\begin{aligned} \beta_{t,R}(f_{v_\lambda w_\mu}|_{T_J}) &= \left\{ \begin{matrix} \alpha_t(p_J(\mu))\,\langle\langle v_\lambda \mid w_\mu \rangle\rangle & \text{if} & p_J(\mu) \in p_J(R \cap P) \\ 0 & \text{else} & \end{matrix} \right\} \\ &= \langle\langle v_\lambda \mid te(R) w_\mu \rangle\rangle . \end{aligned}$$

Therefore we have $\beta_{t,R} = \beta_{te(R)}$. Also by Remark 1.5 every homomorphism is of this form. Therefore $\overline{T_J} = \widehat{T}_J$.

□

Remark. As just stated, the proof of $\overline{T} = \widehat{T}$ is similar to part (c) of the last proof. This shows that $(\overline{T}, \mathbb{F}[\overline{T}] = \mathbb{F}[X \cap P])$ can be identified with the generalized affine toric variety associated to the saturated submonoid $X \cap P$ of P. In particular all statements of Proposition 1.4 are valid.

Conjugation by elements of N induces an action of the Weyl group $\mathcal{W} \cong N/T$ on $\overline{T} = \widehat{T}$, extending the action on T. It is given explicitely by

$$\sigma(te(R)) = (\sigma t)\, e(\sigma R) \quad , \quad \sigma \in \mathcal{W} ,\ t \in T ,\ R \in \mathcal{R}(X) .$$

5.2. The orbits $\overline{G_J}(L(\Lambda)_\Lambda)$ $(J \subseteq I)$ and $\overline{G}(L(\Lambda)_\Lambda)$

Let $\Lambda \in P^+$. The flag variety $G/P_{\{\, i \mid \Lambda(h_i)=0\,\}}$ can be realized as a subvariety of $\mathbb{P}(L(\Lambda))$. The corresponding affine cone, which we call the Kostant cone, is given by the orbit $G(L(\Lambda)_\Lambda) \subseteq L(\Lambda)$. Although $\overline{G}$ does not act on the flag variety, we show that it acts on this cone, i.e., $\overline{G}(L(\Lambda)_\Lambda) = G(L(\Lambda)_\Lambda)$. This is advantageous, because of the rich combinatorial structure of this cone.

Furthermore we show $\overline{G_J}(L(\Lambda)_\Lambda) = G_J(L(\Lambda)_\Lambda)$, for which we need the following Proposition:

Proposition 5.3. *Let $\Lambda \in P^+$ and $\emptyset \neq J \subseteq I$. With the notations of Proposition 2.37 there exists a $\mathbf{g}(A_J)$-submodule of $L(\Lambda)$, which is a orthogonal complement of $L_J(\Lambda)$.*

PROOF. (1) Let V be a $\mathbf{g}$-module isomorphic to a sum of irreducible highest weight modules $L(N)$, $N \in \mathbf{h}^*$. It decomposes in a direct sum

$$V = \bigoplus_{N \in \mathbf{h}^*} V(N) ,$$

where $V(N)$ is the $L(N)$-isotypical component. Let $\langle\langle \mid \rangle\rangle$ be a nondegenerate contravariant symmetric bilinear form on V. We show that different isotypical components are orthogonal:

Let $v_1 \in V(N_1)$, $v_2 \in V(N_2)$ and $N_1 \neq N_2$. Because $V(N_1)$ is isomorphic to a sum of $L(N_1)$-modules, there exist elements $x \in \mathcal{U}(\mathbf{n}^-)$, $\tilde{v}_1 \in V(N_1)_{N_1}$ such that $v_1 = x\tilde{v}_1$. We get

$$\langle\langle v_1 \mid v_2 \rangle\rangle = \langle\langle x\tilde{v}_1 \mid v_2 \rangle\rangle = \langle\langle \underbrace{\tilde{v}_1}_{\in V_{N_1}} \mid \underbrace{x^* v_2}_{\in \bigoplus_{q \in Q_0^+} V_{N_2 - q}} \rangle\rangle .$$

Different weight spaces are orthogonal. If $N_1 > N_2$, or if N_1, N_2 are incomparable, then $\langle\langle v_1 \mid v_2 \rangle\rangle = 0$. Using the symmetry of $\langle\langle \mid \rangle\rangle$, we conclude $\langle\langle v_1 \mid v_2 \rangle\rangle = 0$ for $N_1 < N_2$.

(2) Recall the notations of Proposition 2.37. Due to the orthogonality of different weight spaces, we have

$$L(\Lambda) = \bigoplus_{C \in P(\Lambda)/\sim} V_C \quad \text{where} \quad V_C \perp V_{C'} \text{ for } C \neq C' . \tag{5.4}$$

$V_{[\Lambda]}$ is a direct sum of admissible irreducible highest weight modules of $\mathbf{g}(A_J)$, and $L_J(\Lambda)$ is an isotypical component. Now the proposition follows from (1).

□

Now we show:

Proposition 5.4. *Let $\Lambda \in P^+$. Let $J \subseteq I$. We have:*

(a) $\overline{G_J}(L(\Lambda)_\Lambda) = G_J(L(\Lambda)_\Lambda)$.

(b) $\overline{G}(L(\Lambda)_\Lambda) = G(L(\Lambda)_\Lambda)$.

PROOF. The proof of (b) is similar to the proof of (a). We only show (a) in the nontrivial case $J \neq \emptyset$.

We equip $L_J(\Lambda)$ with the coordinate ring generated by

$$\bigoplus_{\lambda \in P(L_J(\Lambda))} (L(\Lambda)_\lambda)^* \subseteq L_J(\Lambda)^* .$$

Then due to [**K,P 2**], Section 3A, 3B, and Proposition 2.37, the orbit $G_J(L(\Lambda)_\Lambda) \subseteq L_J(\Lambda)$ is Zariski closed in $L_J(\Lambda)$.

Due to Proposition 5.3 there exists a decomposition $L(\Lambda) = L_J(\Lambda) \oplus R$ with $L_J(\Lambda) \perp R$. Using this decomposition, it is easy to see, that the coordinate ring of $L_J(\Lambda)$ is the restriction of the coordinate ring of $L(\Lambda)$. From this decomposition also follows, that $L_J(\Lambda)$ is Zariski closed in $L(\Lambda)$.

For every $v_\Lambda \in L(\Lambda)_\Lambda$ the map

$$\begin{array}{rcl} \Phi : \textit{gr-Adj}\left(\bigoplus_{N \in P^+} L(N)\right) & \to & L(\Lambda) \\ \phi & \mapsto & \phi v_\Lambda \end{array}$$

is Zariski continuous. Therefore we get

$$\overline{G_J}\, v_\Lambda \;=\; \Phi(\overline{G_J}) \;\subseteq\; \overline{\Phi(G_J)} \;\subseteq\; \overline{G_J\,(L(\Lambda)_\Lambda)} \;=\; G_J\,(L(\Lambda)_\Lambda) \;\; .$$

□

5.3. The coordinate rings and closures of $U_J^\pm$, $(U^J)^\pm$ $(J \subseteq I)$

Kac and Peterson showed in [**K,P 2**], Lemma 4.3, that the coordinate ring $\mathbb{F}\,[U^+]$ is a symmetric algebra in $\{\; f_{v_\Lambda\, x v_\Lambda}\,|_{U^+} \mid x \in \mathbf{n}^- \;\}$, where $v_\Lambda \in L(\Lambda)_\Lambda \setminus \{0\}$, $\Lambda \in P^{++}$. From this result follows:

Corollary 5.5. *Let $J \subseteq I$. Let $\Lambda \in F_{I\setminus J} \cap P^+$, and $v_\Lambda \in L(\Lambda)_\Lambda \setminus \{0\}$. Then:*
$\mathbb{F}\,[U_J^+]$ is a symmetric algebra in $\{\, f_{v_\Lambda\, x v_\Lambda}\,|_{U_J^+} \;|\, x \in \mathbf{n}_J^- \,\}$.
$\mathbb{F}\,[U_J^-]$ is a symmetric algebra in $\{\, f_{x v_\Lambda\, v_\Lambda}\,|_{U_J^-} \;|\, x \in \mathbf{n}_J^- \,\}$.

Proof. The case $J = \emptyset$ is trivial. If $J \neq \emptyset$, then G_J, $\mathbb{F}\,[G_J]$ can be identified with $G(A_J)'$, $\mathbb{F}\,[G(A_J)']$. We get the first statement by using Proposition 2.37, and the result of Kac and Peterson of above.

The second statement follows from the first by applying the comorphism of $* : U_J^- \to U_J^+$, which is an isomorphism of algebras $* : \mathbb{F}\,[U_J^+] \to \mathbb{F}\,[U_J^-]$.

□

For $J \subseteq I$ set $(\Delta^J)^\pm := \Delta^\pm \setminus \sum_{j\in J} \mathbb{Z}\alpha_j$, and $(\mathbf{n}^J)^\pm := \bigoplus_{\alpha\in(\Delta^J)^\pm} \mathbf{g}_\alpha$. Write $\mathbf{n}^J := (\mathbf{n}^J)^+$ for short.

Theorem 5.6. *Let $J \subseteq I$. Let $\Lambda \in F_J \cap P^+$, and $v_\Lambda \in L(\Lambda)_\Lambda \setminus \{0\}$. Then:*
The algebra $\mathbb{F}\,[(U^J)^+]$ is generated by $\{\, f_{v_\Lambda\, x v_\Lambda}\,|_{(U^J)^+} \;|\, x \in (\mathbf{n}^J)^- \,\}$.
The algebra $\mathbb{F}\,[(U^J)^-]$ is generated by $\{\, f_{x v_\Lambda\, v_\Lambda}\,|_{(U^J)^-} \;|\, x \in (\mathbf{n}^J)^- \,\}$.

Proof. We only show the first statement. The second follows by applying the comorphism of $* : (U^J)^- \to (U^J)^+$, which is an isomorphism of algebras $* : \mathbb{F}\,[(U^J)^+] \to \mathbb{F}\,[(U^J)^-]$.

The case $J = I$ is trivial, let $J \neq I$. Main parts of the following proof are similar to the proof of the description of $\mathbb{F}\,[U^+]$, given by Kac and Peterson. But we also use this result for the proof.

• Let $h \in \mathbf{h}$. First we show

$$uh \;-\; h \;\in\; \mathbf{n}^J \;\text{ for all }\; u \in U^J \;\; .$$

It is sufficient to show

$$u_1 \exp(x_1) u_1^{-1}\, u_2 \exp(x_2) u_2^{-1} \;\cdots\; u_p \exp(x_p) u_p^{-1}\, h \;-\; h \;\in\; \mathbf{n}^J$$

for all $\beta_1, \ldots, \beta_p \in \Delta^J \cap \Delta^+_{re}$, $x_1 \in \mathbf{g}_{\beta_1}, \ldots, x_p \in \mathbf{g}_{\beta_p}$, $u_1, \ldots, u_p \in U$, $p \in \mathbb{N}$, by induction over p.

Let $\alpha \in \Delta^+$, $\beta \in (\Delta^J)^+$ such that $\alpha + \beta \in \Delta^+$. The definition of $(\Delta^J)^+$ implies $\alpha + \beta \in (\Delta^J)^+$. From this follows $U_\alpha\, \mathbf{n}^J \subseteq \mathbf{n}^J$ for all $\alpha \in \Delta^+_{re}$. Because U is generated by the root groups U_α, $\alpha \in \Delta^+_{re}$, we also get

$$U\, \mathbf{n}^J \;\subseteq\; \mathbf{n}^J \;\; . \tag{5.5}$$

The begin of the induction $p = 1$ follows by using this inclusion:

$$u_1 \exp(x_1) u_1^{-1}\, h \;-\; h \;\;=\;\; \sum_{k=1}^{\infty} \frac{1}{k!}\, [u_1 x_1 [\ldots [u_1 x_1, h] \ldots]] \;\in\; \mathbf{n}^J \;\; .$$

Now the step of the induction from p to $p+1$:

$$\begin{aligned} & u_1 \exp(x_1) u_1^{-1}\, u_2 \exp(x_2) u_2^{-1} \cdots u_{p+1} \exp(x_{p+1}) u_{p+1}^{-1}\, h - h \\ = \; & \underbrace{u_1 \exp(x_1) u_1^{-1} \underbrace{\Big(u_2 \exp(x_2) u_2^{-1} \cdots u_{p+1} \exp(x_{p+1}) u_{p+1}^{-1}\, h - h \Big)}_{\in\, \mathbf{n}^J \textit{ due to the induction assumption}}}_{\in\, \mathbf{n}^J \textit{ due to } (5.5)} \\ & + \underbrace{u_1 \exp(x_1) u_1^{-1}\, h - h}_{\in\, \mathbf{n}^J \textit{ due to } p=1} \in \mathbf{n}^J \; . \end{aligned}$$

- Let $\mathbb{F}[\mathbf{n}^J]$ be the algebra of functions generated by

$$f_x := (x \mid \cdot\,)|_{\mathbf{n}^J} \;, \quad x \in \mathbf{n}_J^- \; .$$

Denote by $\iota : U^J \to U^J$ the inverse map. For $N \in P^+$ denote by $\psi_N : U^J \to \mathbf{n}^J$ the map

$$\psi_N(u) := u\,\nu^{-1}(N) - \nu^{-1}(N) \quad , \quad u \in U^J \; .$$

Choose $v_N \in L(N)_N \setminus \{0\}$. We show

$$\psi_N^*(f_x) = \iota^*\Big(\frac{1}{\langle\langle v_N \mid v_N \rangle\rangle} f_{v_N\, x v_N} |_{U^J} \Big) \; , \tag{5.6}$$

by checking this equality on the elements $u \in U^J$:

$$\begin{aligned} f_x(\psi_N(u)) &= (x \mid u\,\nu^{-1}(N) - \nu^{-1}(N)) = (u^{-1}x \mid \nu^{-1}(N)) \\ &= N\left((u^{-1}x)_0\right) = \frac{\langle\langle v_N \mid (u^{-1}x)_0 v_N \rangle\rangle}{\langle\langle v_N \mid v_N \rangle\rangle} = \frac{\langle\langle v_N \mid (u^{-1}x) v_N \rangle\rangle}{\langle\langle v_N \mid v_N \rangle\rangle} \\ &= \frac{\langle\langle v_N \mid u^{-1}(x(u v_N)) \rangle\rangle}{\langle\langle v_N \mid v_N \rangle\rangle} = \frac{\langle\langle v_N \mid u^{-1}(x v_N) \rangle\rangle}{\langle\langle v_N \mid v_N \rangle\rangle} \; . \end{aligned}$$

Here $(u^{-1}x)_0$ denotes the $\mathbf{g}_0$-homogeneous part of $u^{-1}x$. We used several times the orthogonality of root spaces $\mathbf{g}_\alpha$ and $\mathbf{g}_\beta$, $\alpha + \beta \neq 0$, and the orthogonality of different weight spaces.

Due to [**K,P 2**], Lemma 4.2, the algebra $\mathbb{F}[U]$ is a Hopf algebra. Therefore also $\mathbb{F}[U^J]$ is a Hopf algebra. Because of (5.6) we get $\psi_N^*\left(\mathbb{F}[\mathbf{n}^J]\right) \subseteq \mathbb{F}[U^J]$. Because of the description of $\mathbb{F}[U]$ from Kac and Peterson of above, we have equality at least for $N \in P^{++}$. Now fix an element $\tilde{\Lambda} \in P^{++}$. We have shown

$$\psi_\Lambda^*\left(\mathbb{F}[\mathbf{n}^J]\right) \subseteq \mathbb{F}[U^J] = \psi_{\tilde{\Lambda}}^*\left(\mathbb{F}[\mathbf{n}^J]\right) \; .$$

The theorem is proved, if we show equality. For every $\alpha \in (\Delta^J)^+$ choose $(\,|\,)$-dual bases $(e_\alpha^{(i)})_{i=1,\ldots,m_\alpha} \subseteq \mathbf{g}_\alpha$, $(e_{-\alpha}^{(i)})_{i=1,\ldots,m_\alpha} \subseteq \mathbf{g}_{-\alpha}$. It is sufficient to show

$$\psi_{\tilde{\Lambda}}^*(f_{e_{-\alpha}^{(j)}}) \in \psi_\Lambda^*\left(\mathbb{F}[\mathbf{n}^J]\right) \quad \text{for all} \quad \alpha \in (\Delta^J)^+ \,, \; j = 1, \ldots, m_\alpha \; .$$

For every $u \in U^J$ write the images $\psi_{\tilde{\Lambda}}(u) \in \mathbf{n}^J$, $\psi_\Lambda(u) \in \mathbf{n}^J$ as linear combinations in the basis $(e_\alpha^{(i)})_{\alpha \in (\Delta^J)^+,\, i=1,\ldots,m_\alpha}$ of $\mathbf{n}^J$:

$$\begin{aligned} \psi_{\tilde{\Lambda}}(u) &= \sum_{\alpha \in (\Delta^J)^+} \sum_j \tilde{\phi}_\alpha^{(j)}(u)\, e_\alpha^{(j)} \; , \\ \psi_\Lambda(u) &= \sum_{\alpha \in (\Delta^J)^+} \sum_j \phi_\alpha^{(j)}(u)\, e_\alpha^{(j)} \; , \end{aligned}$$

where

$$\begin{aligned} \tilde{\phi}^{(j)}_\alpha &:= \left(e^{(j)}_{-\alpha} \mid \psi_{\tilde{\Lambda}}(\,\cdot\,)\right) = \psi^*_{\tilde{\Lambda}}(f_{e^{(j)}_{-\alpha}})\ , \\ \phi^{(j)}_\alpha &:= \left(e^{(j)}_{-\alpha} \mid \psi_{\Lambda}(\,\cdot\,)\right) = \psi^*_{\Lambda}(f_{e^{(j)}_{-\alpha}}) \in \psi^*_\Lambda\left(\mathbb{F}\left[\mathbf{n}^J\right]\right)\ . \end{aligned}$$

Because of $[\nu^{-1}(\Lambda), \nu^{-1}(\tilde{\Lambda})] = 0$, we get

$$\begin{aligned} 0 &= u\,[\nu^{-1}(\Lambda), \nu^{-1}(\tilde{\Lambda})] = [u\,\nu^{-1}(\Lambda), u\,\nu^{-1}(\tilde{\Lambda})] \\ &= [\psi_\Lambda(u), \psi_{\tilde{\Lambda}}(u)] + [\nu^{-1}(\Lambda), \psi_{\tilde{\Lambda}}(u)] + [\psi_\Lambda(u), \nu^{-1}(\tilde{\Lambda})] \\ &= \sum_{\alpha\,\in\,(\Delta^J)^+} \sum_j \left(-\left(\alpha \mid \tilde{\Lambda}\right)\phi^{(j)}_\alpha(u) + (\alpha \mid \Lambda)\,\tilde{\phi}^{(j)}_\alpha(u)\right) e^{(j)}_\alpha \\ &\qquad + \sum_{\alpha,\beta\,\in\,(\Delta^J)^+} \sum_{j,k} \phi^{(j)}_\alpha(u)\tilde{\phi}^{(k)}_\beta(u)\,[e^{(j)}_\alpha, e^{(k)}_\beta]\ . \end{aligned}$$

Taking the γ-homogeneous part, $\gamma \in (\Delta^J)^+$, we find

$$\begin{aligned} 0 &= \sum_j \left(-\left(\gamma \mid \tilde{\Lambda}\right)\phi^{(j)}_\gamma + (\gamma \mid \Lambda)\,\tilde{\phi}^{(j)}_\gamma\right) e^{(j)}_\gamma \\ &\qquad + \sum_{\substack{\alpha,\beta\,\in\,(\Delta^J)^+ \\ \alpha+\beta=\gamma}} \sum_{j,k} \phi^{(j)}_\alpha\tilde{\phi}^{(k)}_\beta\,[e^{(j)}_\alpha, e^{(k)}_\beta]\ . \end{aligned} \tag{5.7}$$

An element $\gamma \in (\Delta^J)^+$ is of the form $\gamma = \sum_{i\,\in\,I} p_i\alpha_i$ with $p_i \in \mathbb{N}_0$, and $p_i \neq 0$ for at least one $i \in I \setminus J$. Because Λ is of the form $\Lambda = \sum_{i\,\in\,I\setminus J} m_i\Lambda_i + \Lambda_{rest}$ with $m_i \in \mathbb{N}$, $\Lambda_{rest} \in P_{rest}$, we find

$$(\gamma \mid \Lambda) = \sum_{i\in I,\, j\in I\setminus J} p_i\,m_j\,(\alpha_i \mid \Lambda_j) + 0 = \sum_{i\,\in\,I\setminus J} p_i\,m_i\,\frac{(\alpha_i \mid \alpha_i)}{2} > 0\ .$$

Now we show

$$\tilde{\phi}^{(j)}_\gamma = \psi^*_{\tilde{\Lambda}}(f_{e^{(j)}_{-\alpha}}) \in \psi^*_\Lambda\left(\mathbb{F}\left[\mathbf{n}^J\right]\right) \tag{5.8}$$

by induction over the height of γ. To begin the induction, let $\gamma \in (\Delta^J)^+$ such that $ht(\gamma) = \min\left\{\ ht(\delta) \mid \delta \in (\Delta^J)^+\ \right\}$. Using (5.7) we get

$$0 = \sum_j \left(-\left(\gamma \mid \tilde{\Lambda}\right)\phi^{(j)}_\gamma + (\gamma \mid \Lambda)\,\tilde{\phi}^{(j)}_\gamma\right) e^{(j)}_\gamma + 0\ .$$

Therefore

$$\tilde{\phi}^{(j)}_\gamma = \frac{\left(\gamma \mid \tilde{\Lambda}\right)}{(\gamma \mid \Lambda)}\,\phi^{(j)}_\gamma \in \psi^*_\Lambda\left(\mathbb{F}\left[\mathbf{n}^J\right]\right) \qquad j = 1, \ldots, m_\gamma\ .$$

Let (5.8) be valid for all $\gamma \in (\Delta^J)^+$ with $ht(\gamma) < p$. If there's no $\gamma \in (\Delta^J)^+$ with $ht(\gamma) \geq p$, there is nothing to show. Otherwise let $\gamma \in (\Delta^J)^+$, such that $ht(\gamma) = \min\left\{\ ht(\delta) \mid \delta \in (\Delta^J)^+\ ,\ ht(\delta) \geq p\ \right\}$. Due to (5.7) we have

$$\begin{aligned} &\sum_j (\gamma \mid \Lambda)\,\tilde{\phi}^{(j)}_\gamma e^{(j)}_\gamma \\ &\quad = \sum_j \underbrace{\left(\gamma \mid \tilde{\Lambda}\right)\phi^{(j)}_\gamma\, e^{(j)}_\gamma}_{\in\,\psi^*_\Lambda(\mathbb{F}[\mathbf{n}^J])} - \sum_{\substack{\alpha,\beta\in(\Delta^J)^+ \\ \alpha+\beta=\gamma}} \sum_{j,k} \underbrace{\phi^{(j)}_\alpha\tilde{\phi}^{(k)}_\beta}_{\in\,\psi^*_\Lambda(\mathbb{F}[\mathbf{n}^J])\ (ind.\ ass.)} [e^{(j)}_\alpha, e^{(k)}_\beta]\ . \end{aligned}$$

Because of $(\gamma \mid \Lambda) \neq 0$, we conclude $\tilde{\phi}_\gamma^{(j)} \in \psi_\Lambda^*(\mathbb{F}[\mathbf{n}^J])$, $j = 1, \ldots, m_\gamma$.

□

Theorem 5.7. *Let $J \subseteq I$. We have:*

(1) $U_J^\pm$ *is closed.*

(2) $(U^J)^\pm$ *is closed.*

Proof. If (1) is valid, then due to

$$(U^J)^\pm = \bigcap_{\sigma \in \mathcal{W}_J} \underbrace{\sigma U^\pm \sigma^{-1}}_{closed}$$

also (2). It is sufficient to show (1) only in the case of '−', because the homeomorphism $*$: gr-Adj(...) → gr-Adj(...) maps U_J^- onto U_J^+. Due to Proposition 3.6, the Zariski closure of U_J^- is given by

$$\overline{U_J^-} = \Big\{ \phi \in gr\text{-}Adj\Big(\bigoplus_{\Lambda \in P^+} L(\Lambda)\Big) \;\Big|\; \exists \text{ hom. of alg. } \beta_\phi : \mathbb{F}[U_J^-] \to \mathbb{F}$$
$$\forall\, v, w \in L(\Lambda),\, \Lambda \in P^+ : \quad \beta_\phi(f_{vw}|_{U_J^-}) = \langle\langle v \mid \phi w \rangle\rangle \Big\},$$

and $\phi \in \overline{U_J^-}$ is uniquely determined by β_ϕ.

Let $\phi \in \overline{U_J^-}$. Choose an element $v_\Lambda \in L(\Lambda)_\Lambda \setminus \{0\}$, $\Lambda \in F_{I \setminus J} \cap P^+$. Due to Proposition 5.4, and the Birkhoff decomposition of G_J, we have

$$\overline{U_J^-}\mathbb{F}v_\Lambda \subseteq \overline{G_J}\mathbb{F}v_\Lambda = G_J\mathbb{F}v_\Lambda = U_J^- N_J \mathbb{F}v_\Lambda .$$

Therefore there exist $u \in U_J^-$, $n_\sigma \in N_J$, $c \in \mathbb{F}$, such that

$$\phi v_\Lambda = c\, u\, n_\sigma\, v_\Lambda .$$

The coordinate ring $\mathbb{F}[U_J^-]$ has been described in Corollary 5.5. Applying β_ϕ to the unit of $\mathbb{F}[U_J^-]$, we get

$$1 = \beta_\phi\left(\frac{f_{v_\Lambda v_\Lambda}|_{U_J^-}}{\langle\langle v_\Lambda \mid v_\Lambda \rangle\rangle}\right) = \frac{c\langle\langle u^* v_\Lambda \mid n_\sigma v_\Lambda \rangle\rangle}{\langle\langle v_\Lambda \mid v_\Lambda \rangle\rangle} = \frac{c\langle\langle v_\Lambda \mid n_\sigma v_\Lambda \rangle\rangle}{\langle\langle v_\Lambda \mid v_\Lambda \rangle\rangle} .$$

Therefore $cn_\sigma v_\Lambda = v_\Lambda$. Applying β_ϕ to the generators of $\mathbb{F}[U_J^-]$, we find

$$\beta_\phi\left(f_{xv_\Lambda v_\Lambda}|_{U_J^-}\right) = \beta_u\left(f_{xv_\Lambda v_\Lambda}|_{U_J^-}\right) \quad \text{for all} \quad x \in \mathbf{n}_J^- .$$

We conclude $\beta_\phi = \beta_u$, and $\phi = u \in U_J^-$.

□

5.4. The closures of G_J ($J \subseteq I$) and G

Let $J \subseteq I$. The monoid $\overline{G_J}$ contains G_J and $\overline{T}_J = \widehat{T}_J$. Because G_J and $\widehat{T}_J$ generate the monoid $\widehat{G}_J$, we get

$$\overline{G_J} \supseteq \widehat{G}_J . \tag{5.9}$$

Similarly we get

$$\overline{G} \supseteq \widehat{G} . \tag{5.10}$$

In this chapter we show that we have even $\overline{G_J} = \widehat{G}_J$ and $\overline{G} = \widehat{G}$.

Kac and Peterson introduced in [**K,P 2**], Section 4D, a multiplicative subset θ_Λ, $\Lambda \in P_I^+$, of $\mathbb{F}[G_I]$. Adopting in our context we have:

Proposition 5.8. *For $\Lambda \in P^+$ choose an element $v_\Lambda \in L(\Lambda)_\Lambda \setminus \{0\}$. Then*

$$\theta_\Lambda \;:=\; \frac{1}{\langle\langle v_\Lambda \mid v_\Lambda \rangle\rangle}\,\langle\langle v_\Lambda \mid \cdot\, v_\Lambda \rangle\rangle\,|_{\overline{G}} \;\in\; \mathbb{F}[\overline{G}] \setminus \{0\}$$

is independent of the chosen highest weight vector v_Λ and the chosen nondegenerate contravariant symmetric bilinear form. The map

$$\begin{array}{ccc} (P^+, +) & \to & (\mathbb{F}[\overline{G}] \setminus \{0\}, \cdot) \\ \Lambda & \mapsto & \theta_\Lambda \end{array}$$

is an injective homomorphisms of monoids.

The proof is essentially the same as in [**K,P 2**]. This multiplicative set will be very important for the following considerations. For the convenience of the reader we sketch the

PROOF. The map $P^+ \to \mathbb{F}[\overline{G}] \setminus \{0\}$ is injective, because already the restrictions $\theta_\Lambda\,|_T = e_\Lambda \in \mathbb{F}[T]$, $\Lambda \in P^+$, are different.
Let $\Lambda, N \in P^+$. For $u_\pm \in U^\pm$, $t \in T$ we have

$$(\theta_\Lambda \theta_N)(u_- t u_+) \;=\; (e_\Lambda e_N)(t) \;=\; e_{\Lambda+N}(t) \;=\; \theta_{\Lambda+N}(u_- t u_+)\,.$$

The set $U^- T U^+$ is the principal open set of G associated with $\theta_\Lambda\,|_G$, $\Lambda \in P^+ \cap C$. G is irreducible, because the coordinate ring $\mathbb{F}[G]$ is an integral domain. Therefore $U^- T U^+$ is dense in G. It is also dense in $\overline{G}$, and we conclude $\theta_\Lambda \theta_N = \theta_{\Lambda+N}$. □

We will prove $\overline{G_J} = \widehat{G}_J$ by induction over $|J|$. To prepare the induction step, two theorems will be important. The first theorem relates certain pnc-varieties $D_{\widehat{G}_J}(\theta_\Lambda\,|_{\widehat{G}_J})$ of principal open sets to the pnc-varieties $\widehat{G}_L$, $L \subsetneqq J$. The second theorem gives either a covering of $\overline{G_J}$, or a covering of $\overline{G_J}$ without a point, by certain principal open sets $D_{\overline{G_J}}(\theta_\Lambda\,|_{\overline{G_J}})$.
The following easy proposition states, which of the these pnc-varieties of principal open sets are the same. For $J \subseteq I$ set $P_J^+ := P_J \cap P^+ = \sum_{j \in J} \mathbb{N}_0 \Lambda_j$. Note that the faces of P_J^+ are given by P_L^+, $L \subseteq J$. The relative interior of a face P_L^+ is given by

$$ri\, P_L^+ \;=\; P_L^{++} \;:=\; \sum_{l \in L} \mathbb{N}\,\Lambda_l \quad , \quad L \subseteq J\,.$$

These relative interiors give a partition of P_J^+.

Proposition 5.9. *For $J \subseteq I$ let $T_J \subseteq A_J \subseteq \overline{G_J}$. The following map is an injective homomorphisms of monoids:*

$$\begin{array}{ccc} (P_J^+, +) & \to & (\mathbb{F}[A_J] \setminus \{0\}, \cdot) \\ \Lambda & \mapsto & \theta_\Lambda\,|_{A_J} \end{array}$$

Let $L \subseteq J$ and $\Lambda \in ri\, P_L^+$. The pnc-variety

$$(\,D_{A_J}(\theta_\Lambda\,|_{A_J})\,,\,\mathbb{F}[D_{A_J}(\theta_\Lambda\,|_{A_J})]\,,\,\mathcal{F}_{D_{A_J}(\theta_\Lambda|_{A_J})}\,)$$

can be described in the following way:

$$\begin{aligned} D_{A_J}(\theta_\Lambda|_{A_J}) &= \{\ x \in A_J \mid \theta_{\Lambda'}(x) \neq 0 \ \textit{for all}\ \Lambda' \in P_L^+ \ \} \\ \mathbb{F}[D_{A_J}(\theta_\Lambda|_{A_J})] &= \{\ \theta_{\Lambda'}|_{A_J} \mid \Lambda' \in P_L^+ \ \}^{-1} \mathbb{F}[A_J] \\ \mathcal{F}_{D_{A_J}(\theta_\Lambda|_{A_J})} &= \left\{ \{\ \theta_{\Lambda'}|_{A_J} \mid \Lambda' \in P_L^+ \ \}^{-1} R \ \middle|\ R \in \mathcal{F}_{A_J} \right\} \end{aligned}$$

Remark. We often write only θ_Λ instead of $\theta_\Lambda|_{A_J}$ for short. We write $(D_{A_J}(P_L^+), \mathbb{F}[D_{A_J}(P_L^+)], \mathcal{F}_{D_{A_J}(P_L^+)})$ for this pnc-variety.

PROOF. The homomorphism of monoids is injective, because the restrictions $\theta_\Lambda|_{T_J} = e_\Lambda \in \mathbb{F}[T_J]$, $\Lambda \in P_J^+$, are different.

The inclusion $D_{A_J}(\theta_\Lambda) \supseteq D_{A_J}(P_L^+)$ is obvious. To show the reverse inclusion let $x \in D_{A_J}(\theta_\Lambda)$. It is easy to check, that

$$\{\ \Lambda' \in P_J^+ \ \big|\ \theta_{\Lambda'}(x) \neq 0 \ \}$$

is a face of P_J^+. Because P_L^+ is the smallest face containing Λ, we have $\theta_{\Lambda'}(x) \neq 0$ for all $\Lambda' \in P_L^+$.

Because of $D_{A_J}(\theta_\Lambda) = D_{A_J}(P_L^+)$ and $\frac{f}{(\theta_\Lambda)^n} = \frac{f}{\theta_{n\Lambda}}$, $f \in \mathbb{F}[A_J]$, $n \in \mathbb{N}$, we have an inclusion of algebras $\mathbb{F}[D_{A_J}(\theta_\Lambda)] \subseteq \mathbb{F}[D_{A_J}(P_L^+)]$. To show equality, it is sufficient to show, that for every $\Lambda' \in P_L^+$ there exist elements $f \in \mathbb{F}[A_J]$, $n \in \mathbb{N}$, such that

$$\frac{f}{(\theta_\Lambda)^n} = \frac{1}{\theta_{\Lambda'}} .$$

Because of $\Lambda \in ri\, P_L^+$ and $\Lambda' \in P_L^+$, there exists a positive integer $n \in \mathbb{N}$, such that $n\Lambda - \Lambda' \in P_L^+$. The function $f = \theta_{n\Lambda - \Lambda'}$ satisfies the last equation.

The natural maps $\mathbb{F}[A_J] \to \mathbb{F}[D_{A_J}(\theta_\Lambda)]$ and $\mathbb{F}[A_J] \to \mathbb{F}[D_{A_J}(P_L^+)]$ coincide. Therefore the filters of ideals are the same.

□

Next we want to describe the pnc-varieties $D_{\widehat{G}_J}(P_{J\setminus L}^+)$, where $L \subsetneqq J$. For this we first need the following proposition:

Proposition 5.10. *Let $K \subseteq I$. Then $\left(D_{T_K}(P_K^+), \mathbb{F}[D_{T_K}(P_K^+)], \mathcal{F}_{D_{T_K}(P_K^+)} \right)$ is a variety, and we have*

$$\begin{aligned} D_{T_K}(P_K^+) &= T_K , \\ \mathbb{F}[D_{T_K}(P_K^+)] &= \mathbb{F}[P_K] . \end{aligned}$$

PROOF. Clearly $D_{T_K}(P_K^+) = T_K$. Due to the definition of $\mathbb{F}[D_{T_K}(P_K^+)]$ and Theorem 5.1 we get

$$\mathbb{F}[D_{T_K}(P_K^+)] = \mathbb{F}[p_K(P \cap X) - P_K^+] .$$

Because of $P_K^+ \subseteq X$ we find

$$P_K = P_K^+ - P_K^+ \subseteq p_K(P \cap X) - P_K^+ \subseteq P_K .$$

$\mathbb{F}[P_K]$ is the classical coordinate ring of the torus T_K. The map $T_K \to \operatorname{Specm} \mathbb{F}[P_K]$ is bijective. Therefore also the map $T_K \to F\text{-}\operatorname{Specm} \mathbb{F}[P_K]$ is bijective, and $(D_{T_K}(P_K^+), \mathbb{F}[D_{T_K}(P_K^+)], \mathcal{F}_{D_{T_K}(P_K^+)})$ is a variety.

□

For the next theorem equip the torus $T_{J\setminus L}$ with the variety structure of the last proposition.

Theorem 5.11. *Let* $L \subsetneqq J \subseteq I$. *Set* $(U_J^{\pm})^L := \bigcap_{\sigma \in \mathcal{W}_L} \sigma U_J^{\pm} \sigma^{-1}$. *We have:*

(a) $D_{\widehat{G}_J}(P^+_{J\setminus L}) = (U_J^-)^L \, \widehat{G}_L \, T_{J\setminus L} \, (U_J^+)^L$.

(b) *The multiplication map*

$$m : \; (U_J^-)^L \times \widehat{G}_L \times T_{J\setminus L} \times (U_J^+)^L \;\; \to \;\; D_{\widehat{G}_J}(P^+_{J\setminus L})$$

is bijective. Its comorphism

$$m^* : \; \mathbb{F}\,[D_{\widehat{G}_J}(P^+_{J\setminus L})] \;\; \to \;\; \mathbb{F}\,[(U_J^-)^L] \otimes \mathbb{F}\,[\widehat{G}_L] \otimes \mathbb{F}\,[P_{J\setminus L}] \otimes \mathbb{F}\,[(U_J^+)^L]$$

exists, and is an isomorphism of algebras.

(c) m^{-1} *is a morphism of pnc-varieties.*

Remarks. (1) A decomposition analogous to (b), $J = I$, $L = \emptyset$, the monoid $\widehat{G}_I$ replaced by the group G_I, has been given by Kac and Peterson in [**K,P 2**], Lemma 4.4.

A similar decomposition of the coordinate rings analogous to the second part of (b), $J = I$, has been given by Kashiwara in [**Kas**], Lemma 5.3.4 and Lemma 5.3.5. There coordinate rings are realized as subalgebras of the duals of universal enveloping algebras belonging to certain Lie algebras, and Kashiwara works with these algebras and its spectra. In particular the Kac-Moody group itself isn't available.

(2) We do not know if m^{-1} is an isomorphism, or if it is no isomorphism of pnc-varieties.

PROOF. First we show part (a): Let $\Lambda \in ri\, P^+_{J\setminus L} = P_J^+ \cap F_L$. Due to Theorem 2.41, (b), an element $x \in \widehat{G}_J$ can be written in the form $x = u_- n_\sigma e(R) u_+$, where $u_\pm \in U_J^\pm$, $n_\sigma \in N_J$ and $R \supseteq R(J^\infty)$. We have

$$0 \neq \theta_\Lambda(x) \;=\; \frac{\langle\langle v_\Lambda \mid u_- n_\sigma e(R) u_+ v_\Lambda \rangle\rangle}{\langle\langle v_\Lambda \mid v_\Lambda \rangle\rangle} \;=\; \frac{\langle\langle v_\Lambda \mid n_\sigma e(R) v_\Lambda \rangle\rangle}{\langle\langle v_\Lambda \mid v_\Lambda \rangle\rangle}$$
$$\iff \quad \Lambda \in R \quad \text{and} \quad \sigma\Lambda = \Lambda \; .$$

Because of $\Lambda \in F_L = F_{L^\infty \cup L^0} \subseteq ri\, R(L^\infty)$ this is equivalent to

$$R \supseteq R(L^\infty) \quad \text{and} \quad \sigma \in \mathcal{W}_L \; .$$

Because of $\mathcal{W}_L \subseteq \mathcal{W}_J$ and $R(L^\infty) \supseteq R(J^\infty)$ we get

$$\begin{aligned} D_{\widehat{G}_J}(P^+_{J\setminus L}) &= D_{\widehat{G}_J}(\theta_\Lambda) \\ &= \{\; u_- n_\sigma e(R) u_+ \mid u_\pm \in U_J^\pm,\; n_\sigma \in \mathcal{W}_L T_J,\; R \supseteq R(L^\infty) \;\} \; . \end{aligned}$$

Using $U_J^- = (U_J^-)^L U_L^-$, $U_J^+ = U_L^+ (U_J^+)^L$, and Theorem 2.41, (b), once more, now for $\widehat{G}_L$, (a) follows.

Part (b) and (c) follow directly from Proposition 3.4, (1), we only have to check that the prerequisites are satisfied:

• Clearly $T_{J\setminus L} = D_{T_{J\setminus L}}(P^+_{J\setminus L})$ is a sub-pnc-variety of $D_{\widehat{G}_J}(P^+_{J\setminus L})$. For all $\Lambda \in P^+_{J\setminus L}$ we have $\theta_\Lambda\,|_{(U_J^\pm)^L} = 1_{(U_J^\pm)^L}$. Due to Proposition 2.43, (b), we also have

$\theta_\Lambda\,|_{\widehat{G}_L} = 1_{\widehat{G}_L}$. From this follows, that $(U_J^\pm)^L$ and $\widehat{G}_L$ are sub-pnc-varieties of $D_{\widehat{G}_J}(P^+_{J\setminus L})$.

• The multiplication map is surjective due to Part (a) of this proof. The remaining requirement on m is satisfied, because $(U_J^\mp)^L$, $\widehat{G}_L$, and $T_{J\setminus L}$ contain the unit.

• To show that the comorphism m^* exists, and is surjective, first note: Due to (5.9) we have $G_K \subseteq \widehat{G}_K \subseteq \overline{G_K}$, $K \subseteq I$. Therefore $\mathbb{F}[\widehat{G}_K]$ is isomorphic to $\mathbb{F}[G_K]$ by the restriction map. Due to the Peter and Weyl theorem for $\mathbb{F}[G_K]$, and Proposition 2.37, we conclude

$$\mathbb{F}[\widehat{G}_K] \quad = \quad \operatorname{span}\left\{ f_{vw}\,|_{\widehat{G}_K} \;\middle|\; v,w \in L_K(\Lambda),\ \Lambda \in P_K^+ \right\} \quad . \tag{5.11}$$

Now let $\Lambda \in P^+_{J\setminus L}$. For $N \in P^+$ we choose $\langle\langle\ |\ \rangle\rangle$-dual bases of $L(N)$, by choosing $\langle\langle\ |\ \rangle\rangle$-dual bases

$$(a_{\lambda\, k})_{k=1,\dots\, m_\lambda} \qquad , \qquad (b_{\lambda\, k})_{k=1,\dots\, m_\lambda}$$

of $L(N)_\lambda$ for every $\lambda \in P(N)$. Let $v \in L(N)_\lambda$, $w \in L(N)_\mu$, $\lambda, \mu \in P(N)$. By checking on the elements of $D_{\widehat{G}_J}(P^+_{J\setminus L})$, using Proposition 2.43, (b), we find

$$m^*\left(\frac{f_{vw}}{\theta_\Lambda}\,|_{D_{\widehat{G}_J}(P^+_{J\setminus L})}\right)$$
$$= \sum_{\substack{\lambda' \geq \lambda\,,\, i \\ \mu' \geq \mu\,,\, j}} f_{v\, a_{\lambda' i}}\,|_{(U_J^-)^L} \otimes f_{b_{\lambda' i}\, a_{\mu' j}}\,|_{\widehat{G}_L} \otimes e_{p_{J\setminus L}(\mu')-\Lambda} \otimes f_{b_{\mu' j}\, w}\,|_{(U_J^+)^L} \quad .$$

Since $P(N) \subset N - Q_0^+$ this sum is really finite. Using (5.11) for $K = J$ we find that $m^* : \mathbb{F}[D_{\widehat{G}_J}(P^+_{J\setminus L})] \to \mathbb{F}[(U_J^-)^L] \otimes \mathbb{F}[\widehat{G}_L] \otimes \mathbb{F}[P_{J\setminus L}] \otimes \mathbb{F}[(U_J^+)^L]$ exists. To show the surjectivity of m^*, it is sufficient to find elements of $\mathbb{F}[D_{\widehat{G}_J}(P^+_{J\setminus L})]$, which are mapped by m^* onto a system of generators of the algebra $\mathbb{F}[(U_J^-)^L] \otimes \mathbb{F}[\widehat{G}_L] \otimes \mathbb{F}[P_{J\setminus L}] \otimes \mathbb{F}[(U_J^+)^L]$:

By using Proposition 2.43, (b), it is easy to see, that for all $\Lambda, N \in P^+_{J\setminus L}$ we have

$$m^*\left(\frac{\theta_N}{\theta_\Lambda}\,|_{D_{\widehat{G}_J}(P^+_{J\setminus L})}\right) \quad = \quad 1 \otimes 1 \otimes e_{N-\Lambda} \otimes 1 \quad .$$

Using the same proposition we find for $\Lambda \in ri\, P^+_{J\setminus L} = P^{++}_{J\setminus L} = P^+_J \cap F_L$, $v_\Lambda \in L(\Lambda)_\Lambda \setminus \{0\}$, $x \in (\mathbf{n}_J^-)^L$:

$$m^*\left(\frac{f_{v_\Lambda\, x v_\Lambda}}{\theta_\Lambda}\,|_{D_{\widehat{G}_J}(P^+_{J\setminus L})}\right) \quad = \quad 1 \otimes 1 \otimes 1 \otimes f_{v_\Lambda\, x v_\Lambda}\,|_{(U_J^+)^L} \quad ,$$
$$m^*\left(\frac{f_{x v_\Lambda\, v_\Lambda}}{\theta_\Lambda}\,|_{D_{\widehat{G}_J}(P^+_{J\setminus L})}\right) \quad = \quad f_{x v_\Lambda\, v_\Lambda}\,|_{(U_J^-)^L} \otimes 1 \otimes 1 \otimes 1 \quad .$$

The elements $f_{v_\Lambda\, x v_\Lambda}\,|_{(U_J^+)^L}$, $x \in (\mathbf{n}_J^-)^L$ generate $\mathbb{F}[(U_J^+)^L]$, and the elements $f_{x v_\Lambda\, v_\Lambda}\,|_{(U_J^-)^L}$, $x \in (\mathbf{n}_J^-)^L$ generate $\mathbb{F}[(U_J^-)^L]$. By identifying G_J, $\mathbb{F}[G_J]$ with $G(A_J)'$, $\mathbb{F}[G(A_J)']$, this follows from Theorem 5.6 and Proposition 2.37.

Let $N \in P_L^+$ and $v_\lambda \in L_L(N)_\lambda$, $w_\mu \in L_L(N)_\mu$ where $\lambda, \mu \in P(L_L(N))$. Because the weight μ is of the form $\mu = N - \sum_{l\in L} k_l \alpha_l$ with $k_l \in \mathbb{N}_0$, we have

$$\mu(h_i) \;=\; \underbrace{N(h_i)}_{\geq 0} - \sum_{l\in L} k_l \underbrace{\alpha_l(h_i)}_{\leq 0} \;\geq\; 0 \quad \text{for all} \quad i \in J \setminus L\ .$$

Set $\Lambda := \sum_{i\in J\setminus L} \mu(h_i)\Lambda_i \in P_{J\setminus L}^+$. By using Proposition 2.44 we find

$$m^*\left(\frac{f_{v_\lambda w_\mu}}{\theta_\Lambda}\,|_{D_{\widehat{G}_J}(P_{J\setminus L}^+)}\right) \;=\; 1 \otimes f_{v_\lambda w_\mu}\,|_{\widehat{G}_L} \otimes 1 \otimes 1\ .$$

Due to (5.11) the elements $f_{v_\lambda w_\mu}\,|_{\widehat{G}_L}$ generate $\mathbb{F}\,[\widehat{G}_L]$. □

We use the principal open sets $D_{\overline{G_J}}(P_{J\setminus L}^+)$, $L \subsetneq J$, to give a covering either of $\overline{G_J}$, or of $\overline{G_J}$ without one point. We first give a theorem, which is used to distinguish these two cases, and to describe the vanishing ideal of this point.

Let $\emptyset \neq J \subseteq I$. $\mathbb{F}\,[\overline{G_J}]$ is isomorphic to $\mathbb{F}\,[G_J]$ by the restriction map. Using the Peter and Weyl theorem for $\mathbb{F}\,[G_J]$ and Proposition 2.37, we get an $\overline{G_J}\times\overline{G_J}$-equivariant linear bijective map:

$$\begin{array}{rccc} \overline{\Phi}_J: & \bigoplus\limits_{\Lambda\in P_J^+} L_J(\Lambda)\otimes L_J(\Lambda) & \to & \mathbb{F}\,[\overline{G_J}] \\ & v\otimes w & \mapsto & f_{vw}\,|_{\overline{G_J}} \end{array}$$

Theorem 5.12. *Set* $\mathcal{N}_J := \overline{\Phi}_J\left(\bigoplus_{\Lambda\in P_J^+\setminus\{0\}} L_J(\Lambda)\otimes L_J(\Lambda)\right)$. *Then we have:*
$\mathcal{N}_J = I\Big(e(R(J))\Big)$ *if* $J = J^\infty$.
$\mathcal{N}_J$ *is no ideal if* $J \neq J^\infty$.

PROOF. (a) If $J = J^\infty$ we have to show

$$e(R(J))v \;=\; \begin{cases} v & \text{if} \quad v \in L_J(0) \\ 0 & \text{if} \quad v \in L_J(\Lambda)\,,\ \Lambda \in P_J^+\setminus\{0\} \end{cases}\ .$$

We have $P(L_J(0)) = \{0\} \subseteq R(J)$. It remains to show $P(L_J(\Lambda))\cap R(J) = \emptyset$ for all $\Lambda \in P_J^+\setminus\{0\}$.

Suppose there exist elements $\Lambda \in P_J^+\setminus\{0\}$, $q = \sum_{i\in J} m_i\alpha_i \in (Q_J)_0^+$, $c \in R(J)$ such that $\Lambda - q = c$. Then we get

$$0 \;\leq\; \Lambda(h_j) \;=\; \sum_{i\in J} m_i\,\alpha_i(h_j) + 0 \;=\; \sum_{i\in J} a_{ji}\,m_i \quad \text{for all} \quad j\in J\ .$$

The generalized Cartan matrix A_J has no component of finite type. Due to [**K**], Theorem 4.3, we get

$$\Lambda(h_j) \;=\; \sum_{i\in J} a_{ji}\,m_i \;=\; 0 \quad \text{for all} \quad j \in J\,,$$

which contradicts $\Lambda \neq 0$.

(b) Let $J = J^0$ and suppose $\mathcal{N}_J$ is an ideal of $\mathbb{F}\,[\overline{G_J}]$. The algebra $\mathbb{F}\,[\overline{G_J}]$ is isomorphic to $\mathbb{F}\,[G_J]$ by the restriction map. The map $G_J \to \operatorname{Specm}\mathbb{F}\,[G_J]$ is bijective, because A_J has only components of finite type. Therefore there exists a

$g \in G_J$, such that $\mathcal{N}_J = I(g)$. Choose an element $v \in L_J(\Lambda) \setminus \{0\}$, $\Lambda \in P_J^+ \setminus \{0\}$. Then

$$v = g^{-1} \underbrace{(gv)}_{=0} = 0$$

is a contradiction.

(c) Let $J = J^0 \cup J^\infty$ with $J^0 \neq \emptyset$ and $J^\infty \neq \emptyset$. The generalized Cartan matrix A_J decomposes into the sum $A_J = A_{J^0} \oplus A_{J^\infty}$. Recall that we have $G_J \cong G_{J^0} \times G_{J^\infty}$, and $\mathbb{F}[G_J] \cong \mathbb{F}[G_{J^0}] \otimes \mathbb{F}[G_{J^\infty}]$. The subalgebra corresponding to $\mathbb{F}[G_{J^0}] \otimes 1$ is given by the image of the embedding

$$i : \mathbb{F}[G_{J^0}] \to \mathbb{F}[G_J] \ ,$$

defined by

$$i(f_{vw}|_{G_{J^0}}) := f_{vw}|_{G_J} \quad , \quad v, w \in L_{J^0}(\Lambda) = L_J(\Lambda)\, , \ \Lambda \in P_{J^0}^+ \ .$$

Suppose $\mathcal{N}_J$ is an ideal and let $\phi : \mathbb{F}[G_J] \to \mathbb{F}$ be the corresponding homomorphism of algebras with kernel $\mathcal{N}_J|_{G_J}$. Then $\phi \circ i : \mathbb{F}[G_{J^0}] \to \mathbb{F}$ is a homomorphism of algebras with kernel $\mathcal{N}_{J^0}|_{G_{J^0}}$, which contradicts (b).

□

Let $\sigma, \tau \in \mathcal{W}_J$, and choose elements $n_\sigma, n_\tau \in N_J$ belonging to σ, τ. Let $K \subseteq I$. Since $T_J D_{\overline{G_J}}(P_K^+) T_J = D_{\overline{G_J}}(P_K^+)$, the set

$$\sigma D_{\overline{G_J}}(P_K^+)\tau := n_\sigma D_{\overline{G_J}}(P_K^+) n_\tau$$

is independent of the chosen elements n_σ, n_τ.

Theorem 5.13. *Let $\emptyset \neq J \subseteq I$. Then*

$$\bigcup_{\sigma,\tau \in \mathcal{W}_J} \bigcup_{\emptyset \neq K \subseteq J} \sigma D_{\overline{G_J}}(P_K^+)\tau = \begin{cases} \overline{G_J} & \text{if} \quad J \neq J^\infty \\ \overline{G_J} \setminus \{e(R(J))\} & \text{if} \quad J = J^\infty \end{cases} .$$

Remarks. (1) For $J = J^\infty$ the element $e(R(J))$ is the zero of $\overline{G_J}$. This follows from Proposition 2.42, and $\overline{G_J} = \widehat{G}_J$, which we will prove in the next theorem.

(2) It is easy to see that it is sufficient to take the second union only over the sets $K = \{j\}$, $j \in J$. In general it is not possible to omit the union over the elements of the Weyl group on the left or on the right. This can be checked by using the generalized Cartan matrix of Example 2.6, (3). Choose $J = I$. Not all idempotents of E would be contained in these unions.

(3) A similar decomposition for the full maximal spectrum of $\mathbb{F}[G]$ has been given by Kashiwara in [**Kas**], Proposition 6.3.1.

PROOF. For every $\Lambda \in P_J^+ \setminus \{0\}$ choose an element $v_\Lambda \in L(\Lambda)_\Lambda \setminus \{0\}$.

We have $D_{\overline{G_J}}(P_K^+) = D_{\overline{G_J}}(\theta_\Lambda)$, $\Lambda \in ri\, P_K^+$. Because of $\bigcup_{\emptyset \neq K \subseteq J} ri\, P_K^+ = P_J^+ \setminus \{0\}$ we find:

$$\overline{G_J} \setminus \left(\bigcup_{\emptyset \neq K \subseteq J} \bigcup_{\sigma,\tau \in \mathcal{W}_J} \sigma D_{\overline{G_J}}(P_K^+)\tau \right)$$
$$= \left\{ x \in \overline{G_J} \;\middle|\; \forall \sigma, \tau \in \mathcal{W}_J \ \forall \Lambda \in P_J^+ \setminus \{0\} : \quad \langle\langle n_\sigma v_\Lambda \mid x n_\tau v_\Lambda \rangle\rangle = 0 \right\}$$

If $J = J^\infty$, then due to the last theorem $e(R(J))$ belongs to this set. Now let $\emptyset \neq J \subseteq I$ be arbitrary, and $x \in \overline{G_J}$ an element of this set, i.e.,

$$\langle\langle n_\sigma v_\Lambda \mid x n_\tau v_\Lambda \rangle\rangle \;=\; 0 \quad \text{for all} \quad \sigma, \tau \in \mathcal{W}_J,\ \Lambda \in P_J^+ \setminus \{0\}\ . \tag{5.12}$$

Fix $\Lambda \in P_J^+ \setminus \{0\}$. For an element $v \in L(\Lambda)$ denote by v_λ its component in the weight space $L(\Lambda)_\lambda$, and set $supp(v) := \{\ \lambda \mid v_\lambda \neq 0\ \}$. Denote by $co(supp(v))$ the convex hull of $supp(v)$ in $\mathbf{h}_\mathbb{R}^*$.

Because $\overline{G_J}$ is a monoid containing G_J, we have $x n_\tau \in \overline{G_J}$. Due to Proposition 5.4 we find

$$x n_\tau v_\Lambda \;\in\; \overline{G_J} v_\Lambda \;\subseteq\; G_J L(\Lambda)_\Lambda \;\subseteq\; L_J(\Lambda) \quad \text{for all} \quad \tau \in \mathcal{W}_J\ .$$

Using [**K,P 1**], Lemma 4, the vertices of the convex hull $co(supp(x n_\tau v_\Lambda))$ are elements of $\mathcal{W}_J \Lambda$. The weight spaces $L(\Lambda)_{\sigma\Lambda}$, $\sigma \in \mathcal{W}_J$, are one-dimensional, and different weight spaces are orthogonal. Because of (5.12) there are no vertices of $co(supp(x n_\tau v_\Lambda))$. Therefore we have

$$0 \;=\; x n_\tau v_\Lambda \quad \text{for all} \quad \tau \in \mathcal{W}_J\ .$$

From this follows

$$0 \;=\; \langle\langle g v_\Lambda \mid x n_\tau v_\Lambda \rangle\rangle \;=\; \langle\langle x^* g v_\Lambda \mid n_\tau v_\Lambda \rangle\rangle \quad \text{for all} \quad g \in G_J,\ \tau \in \mathcal{W}_J\ .$$

The involution $*$ is Zariski continuous, leaving G_J invariant. Therefore also $\overline{G_J}$ is invariant, and we get $x^* \in \overline{G_J}$. Now we can use the same argument as above to show $x^* g v_\Lambda = 0$, $g \in G_J$. Since $L_J(\Lambda)$ is spanned by $G_J v_\Lambda$, we find

$$x^* L_J(\Lambda) \;=\; \{0\}\ .$$

Because this equation is valid for all $\Lambda \in P_J^+ \setminus \{0\}$, we get $\mathcal{N}_J \subseteq I\,(x^*)$. We have even equality because both are 1-codimensional subspaces of $\mathbb{F}\,[\overline{G_J}]$.

Due to the last theorem we conclude $J = J^\infty$, and $I\,(\,x^*\,) = I\,(\,e(R(J))\,)$. Because $\mathbb{F}\,[\overline{G_J}]$ separates the points of $\overline{G_J}$, we get $x^* = e(R(J))$. Therefore $x = e(R(J))^* = e(R(J))$.

□

The following theorem is one of the main results of this book:

Theorem 5.14. *Let $J \subseteq I$. We have:*

(1) $\overline{G_J} = \widehat{G}_J\ .$

(2) $\overline{G} = \widehat{G}\ .$

Proof. At the beginning of this section we have already proved the inclusions '$\supseteq$'. We only have to show the reverse inclusions.

To (1). We show $\overline{G_J} \subseteq \widehat{G}_J$ by induction over $|J|$: The case $J = \emptyset$ is trivial. Let $|J| = 1$. Since $J^\infty = \emptyset$ we have $\widehat{G}_J = G_J$. The map $G_J \to \operatorname{Specm} \mathbb{F}\,[G_J]$ is bijective, because $(G_J,\ \mathbb{F}\,[G_J])$ can be identified with $(SL(2,\mathbb{F}),\ \mathbb{F}\,[SL(2,\mathbb{F})])$. Therefore also the map $G_J \to F\text{-}\operatorname{Specm} \mathbb{F}\,[G_J]$ is bijective, and G_J is Zariski closed.

Now the step of the induction from $|J| \leq m$ to $|J| = m+1$, $(1 < m < |I|)$: Let $L \subsetneqq J$. Due to Proposition 5.10 $T_{J \setminus L}$ is a variety. Using the induction assumption we find that $\widehat{G}_L$ is Zariski closed. Therefore $\widehat{G}_L$ is a variety. Left and right multiplications with elements of G are Zariski homeomorphisms. Using Theorem 5.7 we see that the groups

$$(U_J^{\pm})^L \quad = \quad \bigcap_{\sigma \in \mathcal{W}_L} \underbrace{\sigma U_J^{\pm} \sigma^{-1}}_{closed} \ .$$

are Zariski closed. Therefore also $(U_J^{\pm})^L$ are varieties.

Due to Theorem 5.11 there is a bijective morphism

$$m^{-1} : \ D_{\widehat{G}_J}(P^+_{J\setminus L}) \quad \to \quad (U_J^-)^L \times \widehat{G}_L \times T_{J\setminus L} \times (U_J^+)^L \ ,$$

whose comorphism is also bijective. Applying Proposition 3.4, (2), we find that $D_{\widehat{G}_J}(P^+_{J\setminus L})$ is a variety. We have $\widehat{G}_J \subseteq \overline{G_J}$. Applying Proposition 3.5 we get $D_{\overline{G_J}}(P^+_{J\setminus L}) = D_{\widehat{G}_J}(P^+_{J\setminus L})$. Therefore the principal open set $D_{\overline{G_J}}(P^+_{J\setminus L})$ is contained in $\widehat{G}_J$. Using Theorem 5.13 we find

$$\overline{G_J} \;=\; \bigcup_{\sigma,\tau \in \mathcal{W}_J} \bigcup_{L \subsetneqq J} \underbrace{\sigma D_{\overline{G_J}}(P^+_{J\setminus L})\tau}_{\subseteq \widehat{G}_J} \cup \left\{ \begin{array}{ccc} \emptyset & \text{if} & J \neq J^\infty \\ e(R(J)) & \text{if} & J = J^\infty \end{array} \right\} \subseteq \widehat{G}_J \ .$$

To (2). It is easy to check that Proposition 3.4, (1), can be applied to the multiplication map $m : \widehat{G}_I \times T_{rest} \to \widehat{G}$. Therefore m^{-1} is a bijective morphism, whose comorphism is also bijective. Due to part (1) $\widehat{G}_I$ is Zariski closed. Due to Theorem 5.2 T_{rest} is Zariski closed. Therefore $\widehat{G}_I$ and T_{rest} are varieties. Applying Proposition 3.4, (2), we find that $\widehat{G}$ is a variety. Therefore $\widehat{G}$ is Zariski closed. We conclude $\overline{G} \subseteq \widehat{G}$.

□

5.5. The closures of N_J $(J \subseteq I)$ and N

The Weyl monoid $\widehat{\mathcal{W}} \cong \widehat{N}/T$ plays the same role for the Bruhat and Birkhoff decompositions as the Renner monoid for the Bruhat and Birkhoff decompositions of a reductive algebraic monoid. To make this analogy even closer, we show $\overline{N} = \widehat{N}$.

The proof of $\overline{N} = \widehat{N}$ can be regarded as a simplified version of the proof of $\overline{G} = \widehat{G}$. We only state the main steps, some more details can be found in [**M 1**]. By the same methods as in the proof of Theorem 5.11 it is possible to show:

Theorem 5.15. *Let $L \subsetneqq J \subseteq I$. We have:*

(a) $D_{\widehat{N}_J}(P^+_{J\setminus L}) \;=\; \widehat{N}_L T_{J\setminus L}$

(b) *The multiplication map*

$$m : \ \widehat{N}_L \times T_{J\setminus L} \quad \to \quad D_{\widehat{N}_J}(P^+_{J\setminus L})$$

is bijective, and its comorphism

$$m^* : \ \mathbb{F}[D_{\widehat{N}_J}(P^+_{J\setminus L})] \quad \to \quad \mathbb{F}[\widehat{N}_L] \otimes \mathbb{F}[P_{J\setminus L}]$$

exists and is an isomorphism of algebras.

(c) *m^{-1} is a morphism of pnc-varieties.*

We get by intersecting the equation of Theorem 5.13 with $\overline{N_J}$:

Corollary 5.16. *Let $\emptyset \neq J \subseteq I$. Then*

$$\bigcup_{\sigma,\tau \in \mathcal{W}_J} \bigcup_{\emptyset \neq K \subseteq J} \sigma D_{\overline{N_J}}(P_K^+)\tau \;=\; \begin{cases} \overline{N_J} & \text{if } J \neq J^\infty \\ \overline{N_J} \setminus \{e(R(J))\} & \text{if } J = J^\infty \end{cases} .$$

With this theorem and this corollary we can show the following theorem in the same way as Theorem 5.14, the main step an induction over $|J|$:

Theorem 5.17. *Let $J \subseteq I$. We have:*

(1) $\overline{N_J} = \widehat{N}_J$.

(2) $\overline{N} = \widehat{N}$.

5.6. The openness of the unit group

The unit group of a reductive algebraic monoid is principal open. Here we have:

Proposition 5.18. *The unit group G is open in $\overline{G}$.*

PROOF. We have $\overline{G} = \widehat{G}$. In the same way as in the proof of Theorem 5.11, (a), it is possible to show

$$D_{\overline{G}}(\theta_\Lambda) \;=\; U^- T U^+ \quad \text{for every} \quad \Lambda \in P^{++} \; .$$

Due to [**K,P 2**], Corollary 3.1, we have $G_I = \bigcup_{\sigma \in \mathcal{W}} \sigma U^- T_I U^+$. Note also that left multiplications with elements of G are Zariski-homeomorphisms. Therefore

$$G \;=\; G_I T_{rest} \;=\; \bigcup_{\sigma \in \mathcal{W}} \underbrace{\sigma U^- T U^+}_{open}$$

is open in $\overline{G}$.

□

5.7. The Kac-Peterson-Slodowy part of Specm $\mathbb{F}[G]$

Using the characterization $\widehat{G} = \overline{G}$, it is easy to prove the claim of Peterson about the Kac-Peterson-Slodowy part of the $\mathbb{F}$-valued points of the algebra of strongly regular functions.

A function $f \in \mathbb{F}[\widehat{G}]$ induces a function on Specm $\mathbb{F}[\widehat{G}]$, assigning an element $x \in$ Specm $\mathbb{F}[\widehat{G}]$ the value $\tilde{x}(f)$, where $\tilde{x} : \mathbb{F}[\widehat{G}] \to \mathbb{F}$ is the homomorphism of algebras with kernel x. In this way, Specm $\mathbb{F}[\widehat{G}]$ is equipped with a coordinate ring isomorphic to $\mathbb{F}[\widehat{G}]$. Its Zariski topology coincides with the relative topology of the spectrum of $\mathbb{F}[\widehat{G}]$. For a set $M \subseteq$ Specm $\mathbb{F}[\widehat{G}]$ we denote by $\overline{M}^{Specm}$ its Zariski closure.

We identify the elements of $\widehat{G}$ with the corresponding points of Specm $\mathbb{F}[\widehat{G}]$ given by the vanishing ideals. Note that for a set $M \subseteq \widehat{G} = \overline{G}$ we have

$$\overline{M} \subseteq \overline{M}^{Specm} . \tag{5.13}$$

Theorem 5.19. *We have*

$$\widehat{G} \;=\; \bigcup_{m \in \mathbb{N}} \bigcup_{\beta_1,\ldots,\beta_m \in \Delta_{re}} \overline{U_{\beta_1} \cdots U_{\beta_m} T}^{Specm} .$$

PROOF. We first prove the inclusion '$\subseteq$': We have $\widehat{G} = G'\widehat{T}G'$, and $\widehat{T} = \overline{T}$. The derived Kac-Moody group G' is generated by the root groups U_α, $\alpha \in \Delta_{re}$. Therefore, for every element $\widehat{g} \in \widehat{G}$, there exist roots $\beta_1, \ldots, \beta_p, \gamma_1, \ldots, \gamma_q \in \Delta_{re}$, such that

$$\widehat{g} \in U_{\beta_1} \cdots U_{\beta_p}\overline{T}U_{\gamma_1} \cdots U_{\gamma_q} .$$

Because left and right multiplications with elements of $\widehat{G}$ are Zariski continuous on $\widehat{G} = \overline{G}$, we get

$$\widehat{g} \in \overline{U_{\beta_1} \cdots U_{\beta_p}TU_{\gamma_1} \cdots U_{\gamma_q}} .$$

Because of $U_{\beta_1} \cdots U_{\beta_p}TU_{\gamma_1} \cdots U_{\gamma_q} = U_{\beta_1} \cdots U_{\beta_p}U_{\gamma_1} \cdots U_{\gamma_q}T$, and formula (5.13), we find

$$\widehat{g} \in \overline{U_{\beta_1} \cdots U_{\beta_p}U_{\gamma_1} \cdots U_{\gamma_q}T}^{Specm} .$$

Next we prove the inclusion '$\supseteq$' of the theorem. Let $\Lambda \in P^+$. Choose $\langle\langle \mid \rangle\rangle$-dual bases

$$(a_{\lambda i})_{\lambda\in P(\Lambda),\, i=1,\ldots,m_\lambda} \quad , \quad (b_{\lambda i})_{\lambda\in P(\Lambda),\, i=1,\ldots,m_\lambda}$$

of $L(\Lambda)$, by choosing $\langle\langle \mid \rangle\rangle$-dual bases

$$(a_{\lambda i})_{i=1,\ldots,m_\lambda} \quad , \quad (b_{\lambda i})_{i=1,\ldots,m_\lambda}$$

of every weight space $L(\Lambda)_\lambda$, $\lambda \in P(\Lambda)$.

The root groups U_α, $\alpha \in \Delta_{re}$, and the torus T act locally finite on $L(\Lambda)$. Therefore, for every $v \in L(\Lambda)$ the linear space spanned by $U_{\beta_1} \cdots U_{\beta_m}Tv$ is finite dimensional. For a fixed pair (μ, j) we have

$$\langle\langle a_{\lambda i} \mid U_{\beta_1} \cdots U_{\beta_m}Tb_{\mu j}\rangle\rangle = 0$$

for all pairs (λ, i) except finitely many. Similarly, for a fixed pair (λ, i) we have

$$\langle\langle a_{\lambda i} \mid U_{\beta_1} \cdots U_{\beta_m}Tb_{\mu j}\rangle\rangle = \langle\langle TU_{-\beta_m} \cdots U_{-\beta_1}a_{\lambda i} \mid b_{\mu j}\rangle\rangle = 0$$

for all pairs (μ, j) except finitely many.

This also applies to $\tilde{x}(f_{a_{\lambda i}b_{\mu j}})$ for every $x \in \overline{U_{\beta_1} \cdots U_{\beta_m}T}^{Specm}$. Now fix such an element and define

$$\phi_\Lambda \in End(L(\Lambda)) \quad \text{by} \quad \phi_\Lambda b_{\mu j} := \sum_{\lambda i} b_{\lambda i}\, \tilde{x}(f_{a_{\lambda i}b_{\mu j}}) \;,\; \mu \in P(\Lambda),\ j = 1, \ldots, m_\mu \;,$$

$$\psi_\Lambda \in End(L(\Lambda)) \quad \text{by} \quad \psi_\Lambda a_{\lambda i} := \sum_{\mu j} a_{\mu j}\, \tilde{x}(f_{a_{\lambda i}b_{\mu j}}) \;,\; \lambda \in P(\Lambda),\ i = 1, \ldots, m_\lambda \;.$$

It is easy to check, that ϕ_Λ and ψ_Λ are adjoint maps.

The maps ϕ_Λ, $\Lambda \in P^+$, define an element $\phi \in gr\text{-}Adj\left(\bigoplus_{\Lambda\in P^+} L(\Lambda)\right)$. Note that

$$\tilde{x}(f_{a_{\lambda i}b_{\mu j}}) = \langle\langle a_{\lambda i} \mid \phi b_{\mu j}\rangle\rangle \quad \lambda, \mu \in P(\Lambda),\ i = 1, \ldots, m_\lambda,\ j = 1, \ldots, m_\mu,\ \Lambda \in P^+ .$$

From Proposition 3.6 (a) we get $\phi \in \overline{\widehat{G}} = \widehat{G}$. The evaluation homomorphism corresponding to ϕ coincides with $\tilde{x}$. Therefore $\phi = x$.

□

CHAPTER 6

The proof of $Lie(\overline{G}) \cong \mathbf{g}$

In this chapter we show, that the Lie algebra of $\overline{G}$ is isomorphic to the Kac-Moody algebra $\mathbf{g}$. For the proof we use, that the tangent space of $\overline{G}$ at 1 is isomorphic to the tangent space of the principal open set $D_{\overline{G}}(\theta_\Lambda) = U^- D_{\overline{T}}(\theta_\Lambda) U^+$, $\Lambda \in P^{++}$, at 1. To describe this tangent space, we first determine the Lie algebras of $\overline{T}$ and $U^\pm$.

To state the propositions and theorems in an easy way, we identify the Lie algebras of $\overline{T}$, $U^\pm$ and $\overline{G}$ with subalgebras of the Lie algebra of $gr\text{-}Adj\left(\bigoplus_{\Lambda\in P^+} L(\Lambda)\right)$ (via the tangential maps at 1 of the inclusion maps). This last Lie algebra is identified with $gr\text{-}Adj\left(\bigoplus_{\Lambda\in P^+} L(\Lambda)\right)$. We also identify the tangent space $T_1(D_{\overline{T}}(\theta_\Lambda))$ with $T_1(\overline{T})$ (via the tangential map at 1 of the inclusion map). Similarly for $v \in L(\Lambda)$, we identify the tangent space $T_v(\mathcal{V}_\Lambda)$ of the Kostant cone $\mathcal{V}_\Lambda = G(L(\Lambda)_\Lambda)$ with the corresponding subspace of $T_vL(\Lambda)$, which is identified with $L(\Lambda)$. In the proofs however we distinguish carefully between these different possibilities of giving tangent spaces.

The Kac-Moody algebra $\mathbf{g}$ acts faithfully on the sum $\bigoplus_{\Lambda\in P^+} L(\Lambda)$. We identify $\mathbf{g}$ with the corresponding subalgebra of the Lie algebra $gr\text{-}Adj\left(\bigoplus_{\Lambda\in P^+} L(\Lambda)\right)$.

6.1. The Lie algebra of $\overline{T}$

Proposition 6.1. *We have* $Lie(\overline{T}) = \boldsymbol{h}$.

PROOF. Due to Proposition 3.6 the Lie algebra of $\overline{T}$ is given by

$$Lie(\overline{T}) = \left\{ \phi \in gr\text{-}Adj\left(\bigoplus_{\Lambda\in P^+} L(\Lambda)\right) \;\middle|\; \exists\, \epsilon_\phi \in Der_1(\mathbb{F}[T]) \right.$$
$$\left. \forall\, v, w \in L(\Lambda),\, \Lambda \in P^+ : \quad \epsilon_\phi(f_{vw}|_T) = \langle\langle v \mid \phi w\rangle\rangle \right\},$$

and $\phi \in Lie(\overline{T})$ is uniquely determined by ϵ_ϕ. Due to Proposition 5.1 we have $\mathbb{F}[T] = \mathbb{F}[X \cap P]$.

• We first show the inclusion $\mathbf{h} \subseteq Lie(\overline{T})$. For $h \in \mathbf{h}$ we get an element $\epsilon_h \in Der_1(\mathbb{F}[T])$ by

$$\epsilon_h(e_\lambda) := \lambda(h) \quad , \quad \lambda \in X \cap P\,.$$

This derivation has the required properties, because for $v_\lambda \in L(\Lambda)_\lambda$, $w_\mu \in L(\Lambda)_\mu$, λ, $\mu \in P(\Lambda)$, $\Lambda \in P^+$, we have

$$\epsilon_h(f_{v_\lambda w_\mu}|_T) = \epsilon_h\left(\langle\langle v_\lambda \mid w_\mu\rangle\rangle\, e_\mu\right) = \langle\langle v_\lambda \mid w_\mu\rangle\rangle\, \mu(h) = \langle\langle v_\lambda \mid h w_\mu\rangle\rangle\,.$$

• To show the reverse inclusion, we first prove for $\epsilon \in Der_1(\mathbb{F}[T])$ the following statement by induction over $k \in \mathbb{N}$:

Let $\lambda_1, \ldots, \lambda_k \in X \cap P$ and $p_1, \ldots, p_k \in \mathbb{Z}$, such that $p_1\lambda_1 + \ldots + p_k\lambda_k \in X \cap P$. Then we have

$$\epsilon(e_{p_1\lambda_1+\ldots+p_k\lambda_k}) = p_1\epsilon(e_{\lambda_1}) + \ldots + p_k\epsilon(e_{\lambda_k}) \,. \tag{6.1}$$

The begin of the induction $k = 1$: If $p_1 \geq 0$ this follows by using the derivation properties of ϵ, and the monoid properties of $X \cap P$. If $p_1 < 0$ then $(-p_1)\lambda_1 \in X \cap P$, and we have

$$0 = \epsilon(e_0) = \epsilon(e_{p_1\lambda_1}e_{-p_1\lambda_1}) = \epsilon(e_{p_1\lambda_1}) + (-p_1)\epsilon(e_{\lambda_1}) \,.$$

The step of the induction from $k \to k+1$: If $p_1, \ldots, p_k, p_{k+1}$ are nonnegative, then $p_1\lambda_1 + \ldots + p_k\lambda_k \,,\, p_{k+1}\lambda_{k+1} \in X \cap P$, and we have

$$\begin{aligned} \epsilon(e_{p_1\lambda_1+\ldots+p_{k+1}\lambda_{k+1}}) &= \epsilon(e_{p_1\lambda_1+\ldots+p_k\lambda_k}e_{p_{k+1}\lambda_{k+1}}) \\ &= \epsilon(e_{p_1\lambda_1+\ldots+p_k\lambda_k}) + \epsilon(e_{p_{k+1}\lambda_{k+1}}) \,. \end{aligned}$$

Let one of the p_i's be negative, say $p_{k+1} < 0$. Then $(-p_{k+1})\lambda_{k+1} \in X \cap P$. Therefore also

$$p_1\lambda_1 + \ldots + p_k\lambda_k = (p_1\lambda_1 + \ldots + p_{k+1}\lambda_{k+1}) + (-p_{k+1}\lambda_{k+1}) \in X \cap P \,,$$

and we have

$$\begin{aligned} \epsilon(e_{p_1\lambda_1+\ldots+p_k\lambda_k}) &= \epsilon(e_{p_1\lambda_1+\ldots+p_{k+1}\lambda_{k+1}}e_{-p_{k+1}\lambda_{k+1}}) \\ &= \epsilon(e_{p_1\lambda_1+\ldots\, p_{k+1}\lambda_{k+1}}) + \epsilon(e_{-p_{k+1}\lambda_{k+1}}) \,. \end{aligned}$$

In both cases, using the induction assumption and the begin of the induction, we get equation (6.1).

- Now let $\phi \in Lie(\overline{T})$ and set $h := \sum_{i=1}^{2n-l} \epsilon_\phi(e_{\Lambda_i})h_i$. For all $\lambda \in X \cap P$ we find

$$\epsilon_\phi(e_\lambda) = \epsilon_\phi\left(e_{\sum_i \lambda(h_i)\Lambda_i}\right) = \sum_{i=1}^{2n-l} \lambda(h_i)\epsilon_\phi(e_{\Lambda_i}) = \lambda(h) = \epsilon_h(e_\lambda) \,.$$

Therefore $\epsilon_\phi = \epsilon_h$, and $\phi = h \in \mathbf{h}$.

□

6.2. The Lie algebras of U^+ and U^-

To determine the Lie algebras of $U^\pm$ we need the tangent space of the Kostant cone $\mathcal{V}_\Lambda := G(L(\Lambda)_\Lambda)$ at a highest weight vector.

Theorem 6.2. *Let $\Lambda \in P^+$ and $v_\Lambda \in L(\Lambda)_\Lambda$. We have*

$$T_{v_\Lambda}\mathcal{V}_\Lambda = \begin{cases} \mathbf{g}L(\Lambda)_\Lambda & \text{if} \quad v_\Lambda \neq 0 \\ L(\Lambda) & \text{if} \quad v_\Lambda = 0 \end{cases} \,.$$

PROOF. Denote by L_{high} the $L(2\Lambda)$-isotypical component of $L(\Lambda) \otimes L(\Lambda)$. For $\alpha \in \Delta \cup \{0\}$ choose $(\mid)$-dual bases $(e_\alpha^{(i)})_{i=1}^{m_\alpha} \subseteq \mathbf{g}_\alpha$, $(f_\alpha^{(i)})_{i=1}^{m_\alpha} \subseteq \mathbf{g}_{-\alpha}$, such that $f_\alpha^{(i)} = e_{-\alpha}^{(i)}$ for all $\alpha \in \Delta$, $i = 1, \ldots, m_\alpha$.

- Let $v \in L(\Lambda)$. First we show

$$\begin{aligned} T_v\mathcal{V}_\Lambda &= \{\, x \in L(\Lambda) \mid x \otimes v + v \otimes x \in L_{high} \,\} \qquad (6.2) \\ &= \Big\{ x \in L(\Lambda) \,\Big|\, (\Lambda \mid \Lambda)\Big(x \otimes v + v \otimes x\Big) = \qquad (6.3) \\ &\qquad \sum_{\alpha \in \Delta \cup \{0\}} \sum_i \Big(e_\alpha^{(i)}x \otimes f_\alpha^{(i)}v \,+\, e_\alpha^{(i)}v \otimes f_\alpha^{(i)}x\Big) \Big\} \,. \end{aligned}$$

The equality of the two sets on the right follows, because of

$$L_{high} \;=\; \left\{ \; v \in L(\Lambda)\otimes L(\Lambda) \;\mid\; \tilde{\Omega}\, v \;=\; (\Lambda \mid \Lambda)\, v \;\right\} ,$$

where $\tilde{\Omega}$ is given by the formal expression $\sum_{\alpha\in\Delta\cup\{0\}} e_\alpha^{(i)}\otimes f_\alpha^{(i)}$, compare [**K**], Proposition 14.12.

Due to [**K,P 2**], Theorem 2 a), the vanishing ideal $I\,(\mathcal{V}_\Lambda)$ is generated by the functions

$$(\Lambda \mid \Lambda)\, f_w f_{w'} \;-\; \sum_{\alpha\,\in\,\Delta\cup\{0\}} \;\sum_{i=1}^{m_\alpha} f_{(e_\alpha^{(i)})^* w}\, f_{(f_\alpha^{(i)})^* w'} \qquad , \qquad w, w' \,\in\, L(\Lambda) \;.$$

Therefore $x \in T_v \mathcal{V}_\Lambda$ if and only if the corresponding derivation $\delta_x \in Der_v\,(\,\mathbb{F}\,[L(\Lambda)]\,)$ annihilates these functions, which means that for all $w,\, w' \in L(\Lambda)$ we have

$$\begin{aligned} 0 \;=\; &(\Lambda \mid \Lambda)\Big(\langle\langle w \mid x\rangle\rangle\langle\langle v \mid w'\rangle\rangle + \langle\langle w \mid v\rangle\rangle\langle\langle x \mid w'\rangle\rangle\Big) \\ &- \sum_{\alpha\,\in\,\Delta\cup\{0\}}\;\sum_i \Big(\langle\langle w \mid e_\alpha^{(i)} x\rangle\rangle\langle\langle f_\alpha^{(i)} v \mid w'\rangle\rangle \;+\; \langle\langle w \mid e_\alpha^{(i)} v\rangle\rangle\langle\langle f_\alpha^{(i)} x \mid w'\rangle\rangle\Big) \;. \end{aligned}$$

Because $\langle\langle\;\mid\;\rangle\rangle$ is nondegenerate, this is equivalent for x to be an element of the set given in (6.3).

• Now we can show the statement of the theorem. By equation (6.2) we find $T_0\mathcal{V}_\Lambda = L(\Lambda)$. Now let $v_\Lambda \neq 0$. Since $v_\Lambda\otimes v_\Lambda \in L_{high}$, we get

$$(cyv_\Lambda)\otimes v_\Lambda + v_\Lambda\otimes(cyv_\Lambda) \;=\; cy(v_\Lambda\otimes v_\Lambda)\,\in\, L_{high} \qquad \text{for all} \qquad y\in\mathbf{g},\; c\in\mathbb{F}\;.$$

Due to equation (6.2) we have $T_{v_\Lambda}\mathcal{V}_\Lambda \supseteq \mathbf{g}L(\Lambda)_\Lambda$.

To show the reverse inclusion, let $x \in T_{v_\Lambda}\mathcal{V}_\Lambda$. It is sufficient to show that all the homogenous parts $x_\lambda \in L(\Lambda)_\lambda$ of x are contained in $\mathbf{g}L(\Lambda)_\Lambda$. Due to equation (6.3) we have

$$\begin{aligned} &(\Lambda \mid \Lambda) \sum_{\lambda\,\in\,P(\Lambda)} \underbrace{\Big(x_\lambda\otimes v_\Lambda + v_\Lambda\otimes x_\lambda\Big)}_{\in\,\big(L(\Lambda)\otimes L(\Lambda)\big)_{\Lambda+\lambda}} \\ &=\; \sum_{\lambda\,\in\,P(\Lambda)}\;\sum_{\alpha\in\Delta\cup\{0\}}\;\sum_i \underbrace{\Big(e_\alpha^{(i)}x_\lambda\otimes f_\alpha^{(i)}v_\Lambda + e_\alpha^{(i)}v_\Lambda\otimes f_\alpha^{(i)}x_\lambda\Big)}_{\in\,\big(L(\Lambda)\otimes L(\Lambda)\big)_{\Lambda+\lambda}} \;. \end{aligned}$$

We compare the $(\Lambda+\lambda)$-homogenous parts of this equation. Since $f_\alpha^{(i)} = e_{-\alpha}^{(i)}$ for $\alpha\in\Delta$, and $e_\alpha^{(i)}v_\Lambda = 0$ for $\alpha\in\Delta^+$, we find

$$\begin{aligned} &\Big((\Lambda \mid \Lambda) - (\Lambda \mid \lambda)\Big)\Big(x_\lambda\otimes v_\Lambda + v_\Lambda\otimes x_\lambda\Big) \\ &=\; \sum_{\alpha\in\Delta^+}\;\sum_i \Big(e_\alpha^{(i)}x_\lambda\otimes f_\alpha^{(i)}v_\Lambda + f_\alpha^{(i)}v_\Lambda\otimes e_\alpha^{(i)}x_\lambda\Big) \;. \end{aligned}$$

If $(\Lambda\mid\lambda) = (\Lambda\mid\Lambda)$, then due to [**Mo,Pi**], Section 6.2, Proposition 9, we have $\lambda = \Lambda$. Therefore $x_\lambda \in L(\Lambda)_\Lambda$. If $(\Lambda\mid\lambda) \neq (\Lambda\mid\Lambda)$ apply the map $\langle\langle v_\Lambda \mid \cdot\,\rangle\rangle\otimes id$ to the last equation:

$$\Big((\Lambda \mid \Lambda) - (\Lambda \mid \lambda)\Big)\Big(\langle\langle v_\Lambda \mid x_\lambda\rangle\rangle\, v_\Lambda \,+\, \langle\langle v_\Lambda \mid v_\Lambda\rangle\rangle\, x_\lambda\Big)$$
$$= \sum_{\alpha\in\Delta^+}\sum_i \Big(\langle\langle v_\Lambda \mid e_\alpha^{(i)} x_\lambda\rangle\rangle f_\alpha^{(i)} v_\Lambda \,+\, \underbrace{\langle\langle v_\Lambda \mid f_\alpha^{(i)} v_\Lambda\rangle\rangle}_{=0}\, e_\alpha^{(i)} x_\lambda\Big)\ .$$

Solving for the free x_λ gives $x_\lambda \in \mathbf{n}^- v_\Lambda \oplus \mathbb{F}\, v_\Lambda = \mathbf{g}L(\Lambda)_\Lambda$.

□

Theorem 6.3. *We have* $Lie(U^\pm) = \boldsymbol{n}^\pm$.

PROOF. Using the tangential map at 1 of the isomorphism of varieties $* : U^- \to U^+$, we find $Lie(U^+) = Lie(U^-)^*$. Therefore it is sufficient to show the theorem for U^-.

• First we show $Lie(U^-) \supseteq \mathbf{n}^-$. The Lie algebra $\mathbf{n}^-$ is generated by $\mathbf{g}_\alpha$, $\alpha \in \Delta_{re}^-$. Therefore it is sufficient to show, that for an element $x \in \mathbf{g}_\alpha$, $\alpha \in \Delta_{re}^-$, the corresponding derivation $\delta_x \in Der_1\left(\mathbb{F}\left[gr\text{-}Adj\left(\bigoplus_{\Lambda\in P^+} L(\Lambda)\right)\right]\right)$ annihilates the vanishing ideal $I\,(U^-)$. If $f \in I\,(U^-)$, then we have

$$0 \;=\; f(\exp(tx)) \;=\; f(1) + t\delta_x(f) + \mathrm{O}(t^2) \qquad \text{for all} \qquad t \in \mathbb{F}\ .$$

Because of $|\mathbb{F}\,| = \infty$ we get $\delta_x(f) = 0$.

• To show the reverse inclusion let $\Lambda \in P^{++}$, $v_\Lambda \in L(\Lambda)_\Lambda \setminus \{0\}$. The map

$$\begin{array}{rccc} \Phi: & gr\text{-}Adj\left(\bigoplus\limits_{N\in P^+} L(N)\right) & \to & L(\Lambda) \\ & \psi & \mapsto & \psi v_\Lambda \end{array}$$

is a morphism of varieties. Its tangential map at 1 is given by:

$$\begin{array}{rccc} T_1\Phi: & T_1\left(gr\text{-}Adj\left(\bigoplus\limits_{N\in P^+} L(N)\right)\right) & \to & T_{v_\Lambda}L(\Lambda) \\ & \phi & \mapsto & \phi v_\Lambda \end{array}$$

Because of $\Phi(U^-) \subseteq \mathcal{V}_\Lambda$ we have $(T_1\Phi)(T_1U^-) \subseteq T_{v_\Lambda}\mathcal{V}_\Lambda$. Let $\phi \in Lie(U^-)$ and δ_ϕ the corresponding derivation. Due to the last theorem there exist $y \in \mathbf{n}^-$, $c \in \mathbb{F}$, such that

$$\phi v_\Lambda \;=\; y v_\Lambda + c v_\Lambda\ .$$

For $\tilde{y} \in \mathbf{n}^-$ we find

$$\delta_\phi(f_{\tilde{y}v_\Lambda\, v_\Lambda}|_{U^-}) \;=\; \langle\langle \tilde{y}v_\Lambda \mid yv_\Lambda\rangle\rangle + c\,\underbrace{\langle\langle \tilde{y}v_\Lambda \mid v_\Lambda\rangle\rangle}_{=0} \;=\; \delta_y(f_{\tilde{y}v_\Lambda\, v_\Lambda}|_{U^-})\ .$$

Because the functions $f_{\tilde{y}v_\Lambda\, v_\Lambda}|_{U^-}$, $\tilde{y} \in \mathbf{n}^-$ generate the algebra $\mathbb{F}\,[U^-]$, we have $\delta_\phi = \delta_y$. Therefore $\phi = y \in \mathbf{n}^-$.

□

6.3. The Lie algebra of $\overline{G}$

To determine the Lie algebra of $\overline{G}$, we use the following description of the variety of the principal open set $D_{\overline{G}}(\theta_\Lambda) = U^-TU^+$, $\Lambda \in P^{++}$.

Proposition 6.4. *Let* $\left(D_{\overline{T}}(\theta_\Lambda)\,,\, \mathbb{F}\left[D_{\overline{T}}(\theta_\Lambda)\right]\,,\, \mathcal{F}_{D_{\overline{T}}(\theta_\Lambda)}\right)$ *be the variety of the principal open set* $D_{\overline{T}}(\theta_\Lambda)$, $\Lambda \in P^{++}$.

(1) *We have* $D_{\overline{T}}(\theta_\Lambda) = T$ *and* $\mathbb{F}\left[D_{\overline{T}}(\theta_\Lambda)\right] = \mathbb{F}\left[P\right]$.

(2) *We have* $T_1(D_{\overline{T}}(\theta_\Lambda)) = \boldsymbol{h}$. *The derivation* δ_h *corresponding to* $h \in \boldsymbol{h}$ *is given by* $\delta_h(e_\lambda) = \lambda(h)$, $\lambda \in P$.

PROOF. Part (1) can be proved in a similar way as Proposition 5.10. Due to our identifications, and Proposition 6.1, we have $T_1(D_{\overline{T}}(\theta_\Lambda)) = T_1(\overline{T}) = \mathbf{h}$. The derivation $\epsilon_h \in Der_1(\mathbb{F}\left[\overline{T}\right])$ given in the proof of this proposition extends to the derivation $\delta_h \in Der_1(\mathbb{F}\left[D_{\overline{T}}(\theta_\Lambda)\right])$.

□

Theorem 6.5. *Let* $\Lambda \in P^{++}$. *Let* $v_\Lambda \in L(\Lambda)_\Lambda \setminus \{0\}$. *The multiplication map*

$$m:\ U^- \times D_{\overline{T}}(\theta_\Lambda) \times U^+ \quad \to \quad D_{\overline{G}}(\theta_\Lambda)$$

is bijective. Its comorphism

$$m^*:\ \mathbb{F}\left[D_{\overline{G}}(\theta_\Lambda)\right] \quad \to \quad \mathbb{F}\left[U^-\right] \otimes \mathbb{F}\left[P\right] \otimes \mathbb{F}\left[U^+\right]$$

exists, and is an isomorphism of algebras. Furthermore we have

$$m^*\left(\frac{f_{yv_\Lambda\, v_\Lambda}}{\theta_\Lambda}\,|_{D_{\overline{G}}(\theta_\Lambda)}\right) = f_{yv_\Lambda\, v_\Lambda}\,|_{U^-} \otimes 1 \otimes 1 \quad , \quad y \in \boldsymbol{n}^-\,, \tag{6.4}$$

$$m^*\left(\frac{\theta_N}{(\theta_\Lambda)^n}\,|_{D_{\overline{G}}(\theta_\Lambda)}\right) = 1 \otimes e_{N-n\Lambda} \otimes 1 \quad , \quad N \in P^+,\, n \in \mathbb{N}_0\,, \tag{6.5}$$

$$m^*\left(\frac{f_{v_\Lambda\, yv_\Lambda}}{\theta_\Lambda}\,|_{D_{\overline{G}}(\theta_\Lambda)}\right) = 1 \otimes 1 \otimes f_{v_\Lambda\, yv_\Lambda}\,|_{U^+} \quad , \quad y \in \boldsymbol{n}^-\,. \tag{6.6}$$

The map

$$m^{-1}: D_{\overline{G}}(\theta_\Lambda) \to U^- \times D_{\overline{T}}(\theta_\Lambda) \times U^+$$

is a morphism of varieties. Its tangent map

$$T_1(m^{-1}):\, T_1(D_{\overline{G}}(\theta_\Lambda)) \to T_{(1,1,1)}\left(U^- \times D_{\overline{T}}(\theta_\Lambda)) \times U^+\right)$$

is injective.

PROOF. The proof is similar to the proof of Theorem 5.11. The injectivity of $T_1(m^{-1})$ follows from the surjectivity of $(m^{-1})^*$.

□

The following theorem is one of the main results of this book:

Theorem 6.6. *We have* $Lie(\overline{G}) = \mathbf{g}$.

PROOF. (a) First we show $Lie(\overline{G}) \supseteq \mathbf{g}$. It is sufficient to show $\mathbf{g}_\beta \subseteq Lie(\overline{G})$ for all $\beta \in \Delta_{re}$, and $H \subseteq Lie(\overline{G})$. Due to Proposition 3.6 the Lie algebra of $\overline{G}$ is given by

$$Lie(\overline{G}) = \Big\{ \phi \in gr\text{-}Adj\Big(\bigoplus_{\Lambda \in P^+} L(\Lambda) \Big) \;\Big|\; \exists\, \delta_\phi \in Der_1(\mathbb{F}[G]) \\ \forall\, v, w \in L(\Lambda),\, \Lambda \in P^+ : \quad \delta_\phi(f_{vw}\,|_G) = \langle\langle v \mid \phi w \rangle\rangle \Big\} .$$

Let $x_\beta \in \mathbf{g}_\beta$, $\beta \in \Delta_{re}$, and $h \in H$. Due to the Peter and Weyl theorem for $\mathbb{F}[G]$, there exist linear maps

$$\delta_{x_\beta} : \mathbb{F}[G] \to \mathbb{F} \quad \text{with} \quad \delta_{x_\beta}(f_{vw}\,|_G) := \langle\langle v \mid x_\beta\, w \rangle\rangle , \quad v, w \in L(\Lambda), \quad \Lambda \in P^+ ,$$
$$\delta_h : \mathbb{F}[G] \to \mathbb{F} \quad \text{with} \quad \delta_h(f_{vw}\,|_G) := \langle\langle v \mid h\, w \rangle\rangle , \quad v, w \in L(\Lambda), \quad \Lambda \in P^+ .$$

Using the 1-parameter subgroups $exp(tx_\beta)$, $t \in \mathbb{F}$, and $t_h(s)$, $s \in \mathbb{F}^\times$, we can write these maps in the following forms, which show that δ_{x_β} and δ_h are derivations in 1:

$$\delta_{x_\beta}(f) = \frac{d}{dt}\Big|_0 f(exp(tx_\beta)) \quad , \quad \delta_h(f) = \frac{d}{ds}\Big|_1 f(t_h(s)) \quad , \quad f \in \mathbb{F}[G] .$$

(b) To prove the inclusion $Lie(\overline{G}) \subseteq \mathbf{g}$, we use the notations of the last theorem. We identify

$$T_{(1,1,1)}\left(U^- \times D_{\overline{T}}(\theta_\Lambda)) \times U^+\right) \quad \text{with} \quad T_1(U^-) \times T_1(D_{\overline{T}}(\theta_\Lambda)) \times T_1(U^+) .$$

The tangent map $T_1 j : T_1(D_{\overline{G}}(\theta_\Lambda)) \to T_1\overline{G}$ of the inclusion map $j : D_{\overline{G}}(\theta_\Lambda) \to \overline{G}$ is bijective. Therefore the following concatenation of maps

$$T_1(\overline{G}) \overset{(T_1 j)^{-1}}{\longrightarrow} T_1\left(D_{\overline{G}}(\theta_\Lambda)\right) \overset{T_1(m^{-1})}{\longrightarrow} T_1(U^-) \times T_1(D_{\overline{T}}(\theta_\Lambda)) \times T_1(U^+)$$

is injective. Due to Theorem 6.3 we have $T_1(U^\pm) = \mathbf{n}^\pm$. Due to Proposition 6.4 we have $T_1(D_{\overline{T}}(\theta_\Lambda)) = \mathbf{h}$. We also know $\mathbf{g} \subseteq T_1(\overline{G})$. To show equality, it is sufficient to show

$$\left(T_1(m^{-1}) \circ (T_1 j)^{-1}\right)(y + h + x) = (\, y \,,\, h \,,\, x \,) \quad \text{for all} \quad y \in \mathbf{n}^-,\ h \in \mathbf{h},\ x \in \mathbf{n}^+ .$$

To do this, we have to work with the corresponding derivations. Set $z = y + h + x$, and let $\delta_z \in Der_1(\,\mathbb{F}[\overline{G}]\,)$ be the corresponding derivation. Let $\delta_y \in Der_1(\,\mathbb{F}[U^-]\,)$, $\delta_h \in Der_1(\,\mathbb{F}[D_{\overline{T}}(\theta_\Lambda)]\,)$, and $\delta_x \in Der_1(\,\mathbb{F}[U^+]\,)$ be the derivations corresponding to y, h, and x.

Using equation (6.6) of the last theorem, we find for $\tilde{y} \in \mathbf{n}^-$:

$$\Big((T_1(m^{-1}))\,(T_1 j)^{-1}(\delta_z)\Big)(1 \otimes 1 \otimes f_{v_\Lambda\, \tilde{y} v_\Lambda}\,|_{U^+}) = \Big((T_1 j)^{-1}(\delta_z)\Big)\left(\frac{f_{v_\Lambda\, \tilde{y} v_\Lambda}}{\theta_\Lambda}\,|_{D_{\overline{G}}(\theta_\Lambda)}\right)$$
$$= \frac{1}{\theta_\Lambda(1)^2}\Big(\delta_z(f_{v_\Lambda\, \tilde{y} v_\Lambda}\,|_{\overline{G}})\underbrace{\theta_\Lambda(1)}_{=1} - \underbrace{f_{v_\Lambda\, \tilde{y} v_\Lambda}(1)}_{=0}\,\delta_z(\theta_\Lambda)\Big) = \delta_x(f_{v_\Lambda\, \tilde{y} v_\Lambda}\,|_{U^+}) .$$

Because the algebra $\mathbb{F}[U^+]$ is generated by $f_{v_\Lambda\, \tilde{y} v_\Lambda}\,|_{U^+}$, $\tilde{y} \in \mathbf{n}^-$, we get

$$\left(T_1(m^{-1})\right)(T_1 j)^{-1}(\delta_z) = (\,\ldots\,,\,\ldots\,,\,\delta_x\,) .$$

Similarly we have $\big(T_1(m^{-1})\big)\,(T_1 j)^{-1}(\delta_z) \;=\; (\,\delta_y\,,\,\ldots\,,\,\ldots)\,$. Let $N \in P^+$ and $n \in \mathbb{N}_0$. Using equation (6.5) of the last theorem we find

$$\Big(\big(T_1(m^{-1})\big)\,(T_1 j)^{-1})(\delta_z)\Big)(1 \otimes e_{N-n\Lambda} \otimes 1) \;=\; \Big((T_1 j)^{-1}(\delta_z)\Big)\left(\frac{\theta_N}{(\theta_\Lambda)^n}\,|_{D_{\overline{G}}(\theta_\Lambda)}\right)$$

$$= \frac{1}{\theta_{n\Lambda}(1)^2}\Big(\,\delta_z(\theta_N)\theta_{n\Lambda}(1) - \delta_z(\theta_{n\Lambda})\theta_N(1)\,\Big) \;=\; N(h) - n\Lambda(h) \;=\; \delta_h(e_{N-n\Lambda})\,.$$

Because $\mathbb{F}\,[P]$ is spanned by $e_{N-n\Lambda}$, $N \in P^+$, $n \in \mathbb{N}_0$, we conclude

$$\big(T_1(m^{-1})\big)\,(T_1 j)^{-1}(\delta_z) \;=\; (\,\ldots\,,\,\delta_h\,,\,\ldots)\,.$$

□

Bibliography

[F,F] A. J. Feingold, I. B. Frenkel, A Hyperbolic Kac-Moody Algebra and the Theory of Siegel Modular Forms of Genus 2, Math. Ann. **263** (1983), 87-144.

[Fu] W. Fulton, Introduction to Toric Varieties, Annals of Mathematical Studies **131**, Princeton Univ. Press, Princeton, New Jersey, 1993.

[How] J. M. Howie, An Introduction to Semigroup Theory, London Math. Soc. Monograph **8**, Academic Press, 1976.

[Hu] J. E. Humphreys, Reflection Groups and Coxeter Groups, Cambridge Studies in Advanced Mathematics **29**, Cambridge Univ. Press, 1990.

[K] V. G. Kac, Infinite dimensional Lie algebras, Cambridge Univ. Press, 1990.

[K,P 1] V. G. Kac, D. H. Peterson, Infinite flag varieties and conjugacy theorems, Proc. Natl. Acad. Sci. U.S.A. **80** (1983), 1778-1782.

[K,P 2] V. G. Kac, D. H. Peterson, Regular functions on certain infinite-dimensional groups, in: Arithmetic and Geometry, Progr. in Math. **36**, Birkhäuser, Boston, 1983, 141-166.

[K,P 3] V. G. Kac, D. H. Peterson, Defining relations on certain infinite-dimensional groups, Proceedings of the Cartan conference, Lyon 1984, Astérisque, 1985, Numero hors serie, 165-208.

[Kas] M. Kashiwara, The Flag Manifold of Kac-Moody Lie Algebra, Amer. J. of Math. **111** (1989), Supplement Volume: Algebraic Analysis, Geometry and Number Theory, Ed. J.-i. Jgusa, John Hopkins Univ., Baltimore, 1989, 161-190.

[Loo] E. J. Looijenga, Invariant theory for generalized root systems, Invent. Math. **61** (1980), 1-32.

[M 1] C. Mokler, Die Monoidvervollständigung einer Kac-Moody-Gruppe, Dissertation, Hamburg.

[M 2] C. Mokler, The $\mathbb{F}$-valued points of the algebra of strongly regular functions of a Kac-Moody group, Transformation Groups **7** (2002), 343-378.

[M 3] C. Mokler, Extending the Bruhat order and the length function from the Weyl group to the Weyl monoid, to appear in the J. of Algebra.

[M 4] C. Mokler, The monoid associated to the category of admissible highest weight modules and its category of full duals, in preparation.

[Mo,Pi] R. V. Moody, A. Pianzola, Lie Algebras With Triangular Decompositions, Canad. Math. Soc. Ser. of Monographs and Advanced Texts, John Wiley & Sons, 1995.

[Ne] K.-H. Neeb, Toric varieties and algebraic monoids, Seminar Sophus Lie **2** (1992), 159-187.

[P] D. H. Peterson, letter to P. Slodowy, 1984.

[Pu] M. S. Putcha, Linear Algebraic Monoids, London Math. Soc., Lecture Note Ser. **133**, Cambridge Univ. Press, 1988.

[Re 1] L. E. Renner, Classification of semisimple algebraic monoids, Trans. Amer. Math. Soc. **292** (1985), 193-223.

[Re 2] L. E. Renner, Linear Algebraic Monoids, Encyclopaedia of Mathematical Sciences, Springer Verlag, to appear.

[Ro] R. T. Rockafellar, Convex Analysis, Princeton Univ. Press, Princeton, New Jersey, 1970.

[Sl 1] P. Slodowy, Singularitäten, Kac-Moody-Liealgebren, assoziierte Gruppen und Verallgemeinerungen, Habilitationsschrift, Bonn, 1984.

[Sl 2] P. Slodowy, An adjoint quotient for certain groups attached to Kac-Moody algebras, in "Infinite-dimensional groups with applications", MSRI-Publ. Vol. **4**, Springer, 1985, 307-334.

[Ti 1] J. Tits, Uniqueness and Presentation of Kac-Moody Groups over fields, J. of Algebra **105** (1987), 542-573.

[Ti 2] J. Tits, Twin buildings and groups of Kac-Moody type, London Math. Soc., Lecture Note Ser. **165** (Proceedings of a conference on groups, combinatorics and geometry, Durham 1990), Cambridge Univ. Press, 1992, 249-286.

[Vi] E. B. Vinberg, On Reductive Algebraic Semigroups, Amer. Math. Soc. Transl. (2) **196** (1995), 145-182.

Editorial Information

To be published in the *Memoirs*, a paper must be correct, new, nontrivial, and significant. Further, it must be well written and of interest to a substantial number of mathematicians. Piecemeal results, such as an inconclusive step toward an unproved major theorem or a minor variation on a known result, are in general not acceptable for publication. Papers appearing in *Memoirs* are generally longer than those appearing in *Transactions*, which shares the same editorial committee.

As of November 30, 2004, the backlog for this journal was approximately 5 volumes. This estimate is the result of dividing the number of manuscripts for this journal in the Providence office that have not yet gone to the printer on the above date by the average number of monographs per volume over the previous twelve months, reduced by the number of volumes published in four months (the time necessary for preparing a volume for the printer). (There are 6 volumes per year, each containing at least 4 numbers.)

A Consent to Publish and Copyright Agreement is required before a paper will be published in the *Memoirs*. After a paper is accepted for publication, the Providence office will send a Consent to Publish and Copyright Agreement to all authors of the paper. By submitting a paper to the *Memoirs*, authors certify that the results have not been submitted to nor are they under consideration for publication by another journal, conference proceedings, or similar publication.

Information for Authors

Memoirs are printed from camera copy fully prepared by the author. This means that the finished book will look exactly like the copy submitted.

The paper must contain a *descriptive title* and an *abstract* that summarizes the article in language suitable for workers in the general field (algebra, analysis, etc.). The *descriptive title* should be short, but informative; useless or vague phrases such as "some remarks about" or "concerning" should be avoided. The *abstract* should be at least one complete sentence, and at most 300 words. Included with the footnotes to the paper should be the 2000 *Mathematics Subject Classification* representing the primary and secondary subjects of the article. The classifications are accessible from `www.ams.org/msc/`. The list of classifications is also available in print starting with the 1999 annual index of *Mathematical Reviews*. The Mathematics Subject Classification footnote may be followed by a list of *key words and phrases* describing the subject matter of the article and taken from it. Journal abbreviations used in bibliographies are listed in the latest *Mathematical Reviews* annual index. The series abbreviations are also accessible from `www.ams.org/publications/`. To help in preparing and verifying references, the AMS offers MR Lookup, a Reference Tool for Linking, at `www.ams.org/mrlookup/`. When the manuscript is submitted, authors should supply the editor with electronic addresses if available. These will be printed after the postal address at the end of the article.

Electronically prepared manuscripts. The AMS encourages electronically prepared manuscripts, with a strong preference for $\mathcal{AMS}$-LaTeX. To this end, the Society has prepared $\mathcal{AMS}$-LaTeX author packages for each AMS publication. Author packages include instructions for preparing electronic manuscripts, the *AMS Author Handbook*, samples, and a style file that generates the particular design specifications of that publication series. Though $\mathcal{AMS}$-LaTeX is the highly preferred format of TeX, author packages are also available in $\mathcal{AMS}$-TeX.

Authors may retrieve an author package from e-MATH starting from `www.ams.org/tex/` or via FTP to `ftp.ams.org` (login as `anonymous`, enter username as password, and type `cd pub/author-info`). The *AMS Author Handbook* and the *Instruction Manual* are available in PDF format following the author packages link from `www.ams.org/tex/`. The author package can be obtained free of charge by sending email to `pub@ams.org` (Internet) or from the Publication Division, American Mathematical Society, 201 Charles St., Providence, RI 02904, USA. When requesting an author package, please specify AMS-LaTeX or AMS-TeX, Macintosh or IBM (3.5) format, and the publication in which your paper will appear. Please be sure to include your complete mailing address.

Sending electronic files. After acceptance, the source file(s) should be sent to the Providence office (this includes any TeX source file, any graphics files, and the DVI or PostScript file).

Before sending the source file, be sure you have proofread your paper carefully. The files you send must be the EXACT files used to generate the proof copy that was accepted for publication. For all publications, authors are required to send a printed copy of their paper, which exactly matches the copy approved for publication, along with any graphics that will appear in the paper.

TeX files may be submitted by email, FTP, or on diskette. The DVI file(s) and PostScript files should be submitted only by FTP or on diskette unless they are encoded properly to submit through email. (DVI files are binary and PostScript files tend to be very large.)

Electronically prepared manuscripts can be sent via email to `pub-submit@ams.org` (Internet). The subject line of the message should include the publication code to identify it as a Memoir. TeX source files, DVI files, and PostScript files can be transferred over the Internet by FTP to the Internet node `e-math.ams.org` (130.44.1.100).

Electronic graphics. Comprehensive instructions on preparing graphics are available at `www.ams.org/jourhtml/graphics.html`. A few of the major requirements are given here.

Submit files for graphics as EPS (Encapsulated PostScript) files. This includes graphics originated via a graphics application as well as scanned photographs or other computer-generated images. If this is not possible, TIFF files are acceptable as long as they can be opened in Adobe Photoshop or Illustrator. No matter what method was used to produce the graphic, it is necessary to provide a paper copy to the AMS.

Authors using graphics packages for the creation of electronic art should also avoid the use of any lines thinner than 0.5 points in width. Many graphics packages allow the user to specify a "hairline" for a very thin line. Hairlines often look acceptable when proofed on a typical laser printer. However, when produced on a high-resolution laser imagesetter, hairlines become nearly invisible and will be lost entirely in the final printing process.

Screens should be set to values between 15% and 85%. Screens which fall outside of this range are too light or too dark to print correctly. Variations of screens within a graphic should be no less than 10%.

Inquiries. Any inquiries concerning a paper that has been accepted for publication should be sent directly to the Electronic Prepress Department, American Mathematical Society, 201 Charles St., Providence, RI 02904, USA.

Editors

Titles in This Series

824 **Luen-Chau Li and Serge Parmentier,** On dynamical Poisson groupoids I, 2005

823 **Claus Mokler,** An analogue of a reductive algebraic monoid whose unit group is a Kac-Moody group, 2005

822 **Stefano Pigola, Marco Rigoli, and Alberto G. Setti,** Maximum principles on Riemannian manifolds and applications, 2005

821 **Nicole Bopp and Hubert Rubenthaler,** Local zeta functions attached to the minimal spherical series for a class of symmetric spaces, 2005

820 **Vadim A. Kaimanovich and Mikhail Lyubich,** Conformal and harmonic measures on laminations associated with rational maps, 2005

819 **F. Andreatta and E. Z. Goren,** Hilbert modular forms: Mod p and p-adic aspects, 2005

818 **Tom De Medts,** An algebraic structure for Moufang quadrangles, 2005

817 **Javier Fernández de Bobadilla,** Moduli spaces of polynomials in two variables, 2005

816 **Francis Clarke,** Necessary conditions in dynamic optimization, 2005

815 **Martin Bendersky and Donald M. Davis,** V_1-periodic homotopy groups of $SO(n)$, 2004

814 **Johannes Huebschmann,** Kähler spaces, nilpotent orbits, and singular reduction, 2004

813 **Jeff Groah and Blake Temple,** Shock-wave solutions of the Einstein equations with perfect fluid sources: Existence and consistency by a locally inertial Glimm scheme, 2004

812 **Richard D. Canary and Darryl McCullough,** Homotopy equivalences of 3-manifolds and deformation theory of Kleinian groups, 2004

811 **Ottmar Loos and Erhard Neher,** Locally finite root systems, 2004

810 **W. N. Everitt and L. Markus,** Infinite dimensional complex symplectic spaces, 2004

809 **J. T. Cox, D. A. Dawson, and A. Greven,** Mutually catalytic super branching random walks: Large finite systems and renormalization analysis, 2004

808 **Hagen Meltzer,** Exceptional vector bundles, tilting sheaves and tilting complexes for weighted projective lines, 2004

807 **Carlos A. Cabrelli, Christopher Heil, and Ursula M. Molter,** Self-similarity and multiwavelets in higher dimensions, 2004

806 **Spiros A. Argyros and Andreas Tolias,** Methods in the theory of hereditarily indecomposable Banach spaces, 2004

805 **Philip L. Bowers and Kenneth Stephenson,** Uniformizing dessins and Belyĭ maps via circle packing, 2004

804 **A. Yu Ol'shanskii and M. V. Sapir,** The conjugacy problem and Higman embeddings, 2004

803 **Michael Field and Matthew Nicol,** Ergodic theory of equivariant diffeomorphisms: Markov partitions and stable ergodicity, 2004

802 **Martin W. Liebeck and Gary M. Seitz,** The maximal subgroups of positive dimension in exceptional algebraic groups, 2004

801 **Fabio Ancona and Andrea Marson,** Well-posedness for general 2×2 systems of conservation law, 2004

800 **V. Poénaru and C. Tanas,** Equivariant, almost-arborescent representation of open simply-connected 3-manifolds; A finiteness result, 2004

799 **Barry Mazur and Karl Rubin,** Kolyvagin systems, 2004

798 **Benoît Mselati,** Classification and probabilistic representation of the positive solutions of a semilinear elliptic equation, 2004

797 **Ola Bratteli, Palle E. T. Jorgensen, and Vasyl' Ostrovs'kyĭ,** Representation theory and numerical AF-invariants, 2004

796 **Marc A. Rieffel,** Gromov-Hausdorff distance for quantum metric spaces/Matrix algebras converge to the sphere for quantum Gromov-Hausdorff distance, 2004